Everyday Mathematics®

Skills Link™

Everyday Mathematics®

Cumulative Practice Sets

Columbus, Ohio
Chicago, Illinois

Photo Credits
Cover: Bill Burlingham/Photography
Photo Collage: Herman Adler Design

www.sra4kids.com

SRA/McGraw-Hill

*A Division of The **McGraw-Hill** Companies*

Send all inquiries to:
SRA/McGraw-Hill
P.O. Box 812960
Chicago, IL 60681

Printed in the United States of America.

ISBN 1-57039-966-2

9 10 VHG 07 06 05 04

Contents

Practice Set 1

Write all of your answers on a separate sheet of paper.

For each Fact Minute below, do as many problems as you can in that minute. You can ask someone to time you.

Fact Minute 1

1. $8 + 7$

2. $3 + 1$

3. $9 - 3$

4. $13 - 9$

5. $8 - 4$

6. $6 + 2$

7. $12 - 8$

8. $8 + 5$

9. $10 - 2$

10. $9 - 0$

11. $6 + 6$

12. $7 + 7$

13. $5 + 8$

14. $8 - 3$

15. $18 - 9$

Fact Minute 2

16. $9 + 1$

17. $11 - 4$

18. $7 + 2$

19. $5 + 6$

20. $1 + 7$

21. $9 - 5$

22. $7 + 3$

23. $8 - 4$

24. $6 + 5$

25. $9 + 1$

26. $14 - 6$

27. $4 + 8$

28. $7 + 4$

29. $14 - 2$

30. $10 + 4$

Fact Minute 3

31. $9 - 6$

32. $16 - 8$

33. $4 + 5$

34. $8 + 3$

35. $2 + 5$

36. $15 - 9$

37. $9 - 7$

38. $10 + 1$

39. $7 - 7$

40. $8 + 6$

41. $5 + 5$

42. $8 - 5$

43. $2 + 9$

44. $16 - 6$

45. $14 - 9$

Practice Set 2

SRB 27 77 131

Write all of your answers on a separate sheet of paper.

Match each description with the correct example. Write the letter that identifies the example.

1. point *L*

2. ray *LM*

3. line *LM*

4. line segment *LM*

A. 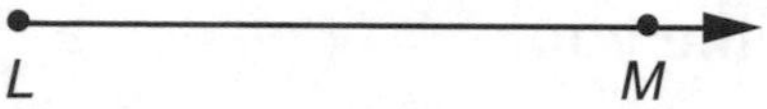

B. 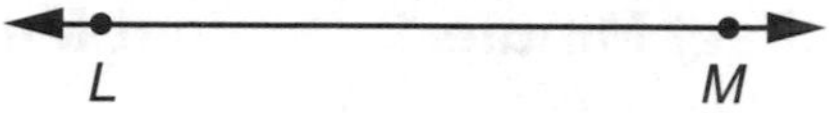

C. 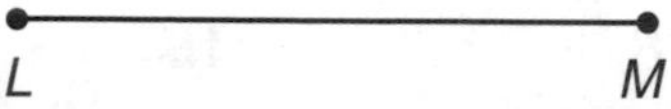

D. 

---

5. Copy the place-value puzzle on your paper. Then use the clues to complete the puzzle.

1,000s	100s	10s	1s

- Write the result of $21 \div 7$ in the ones place.
- Multiply 8×9. Subtract 65. Write the result in the tens place.
- Double the number in the ones place. Write the result in the thousands place.
- Divide 18 by 6. Add 5 and write the result in the hundreds place.

Make a name-collection box for each number listed below. Use as many different numbers and operations as you can.

Example

19
$(6 \times 3) + 1$
$38 \div 2$
$(40 - 25) + 4$

6. 38

7. 7

8. 218

Practice Set 2 (cont.)

Write all of your answers on a separate sheet of paper.

Complete the "What's My Rule?" tables.

9. Rule: out = in × 3

in	out
3	9
4	12
7	
11	
15	

10. Rule:

in	out
8	4
14	7
	9
24	
36	18

Solve.

11. The Coffee-to-Go Cafe uses about 5 gallons of milk per day.

a. About how many gallons of milk does it use in a week (7 days)?

b. How many gallons in 5 weeks?

c. How many in one year (52 weeks)?

12. $18 - 17$

13. $68 - 34$

14. $74 + 27$

15. $5{,}101 - 540$

16. $500 - 290$

17. $402 + 293$

18. $3{,}418 + 6{,}583$

19. $49 - 6$

20. $120 - 30$

21. $81 + 40$

22. $35 - 22$

23. $350 - 150$

Practice Set 3

Write all of your answers on a separate sheet of paper.

Match each description with the correct figure. Write the letter that identifies the figure.

1. rhombus **A.**

2. trapezoid **B.**

3. square **C.**

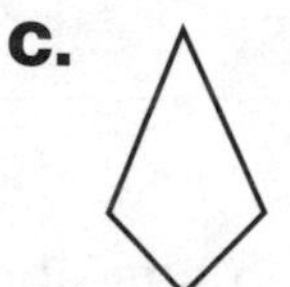

4. kite **D.**

Solve.

5. 2×4

6. $14 - 7$

7. $84 - 27$

8. $300 + 500$

9. $35 - 19$

10. $30 + 83$

11. $43 + 21$

12. 9×0

13. $(50 + 20) \times 4$

14. $27 - (5 \times 4)$

15. $200 + 150 + 100$

16. $500 + 440 + 120$

Practice Set 3 (cont.)

Write all of your answers on a separate sheet of paper.

Complete the frames-and-arrows problems.

Example (with one rule)

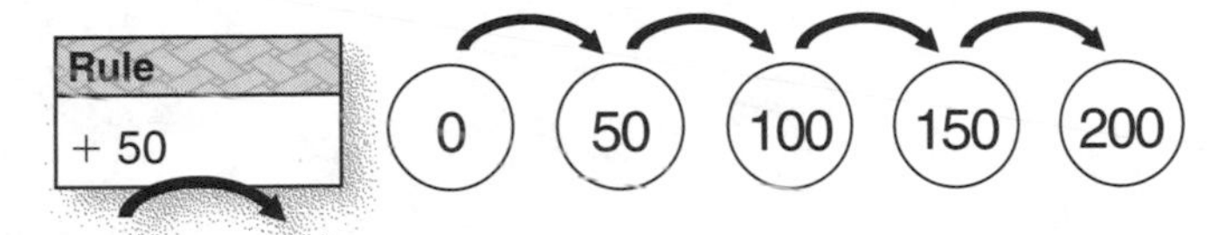

Example (with two rules)

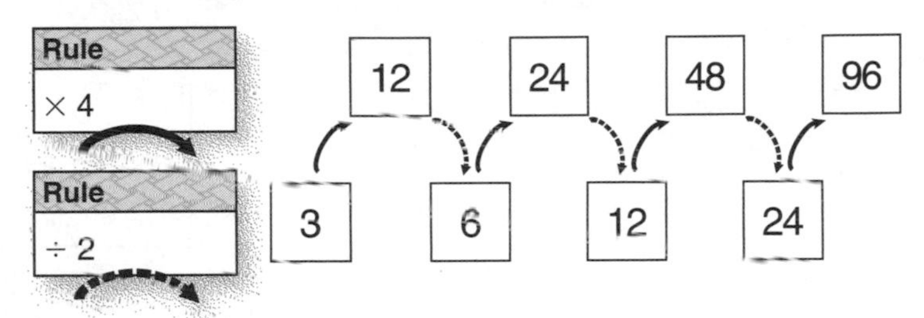

17.

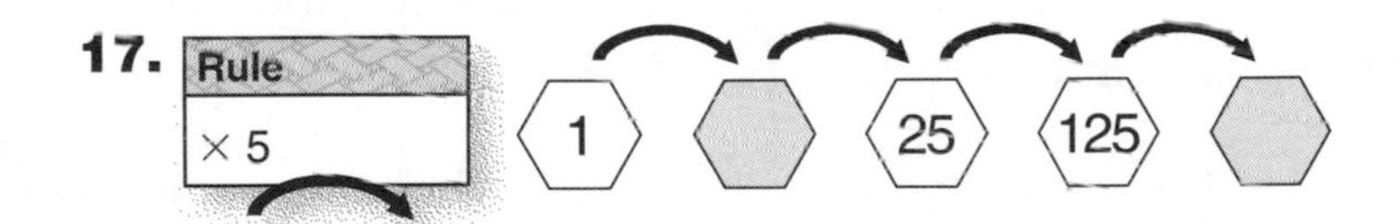

18.

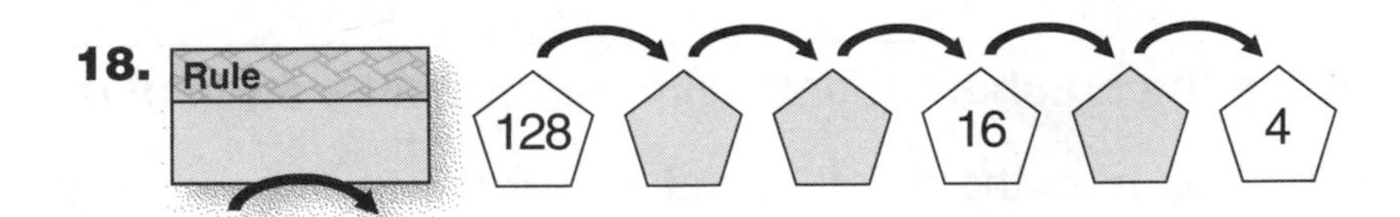

19.

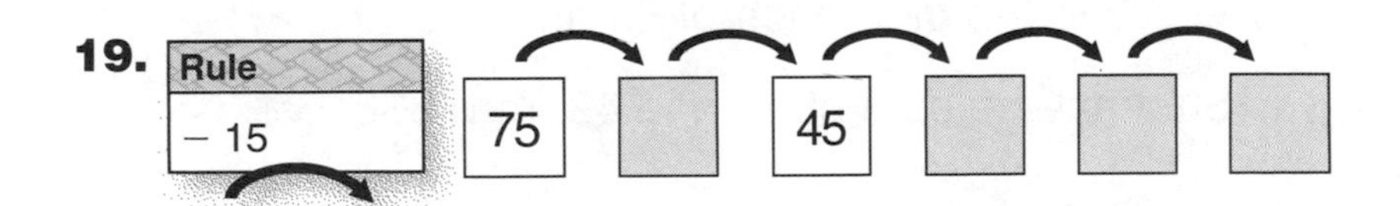

Practice Set 4

Write all of your answers on a separate sheet of paper.

1. Which figures are polygons?

A. 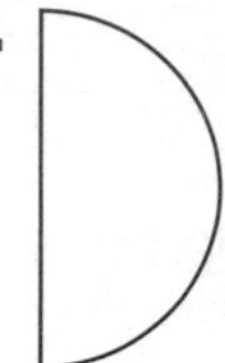

B. 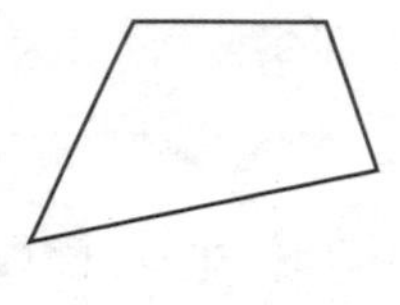

C. 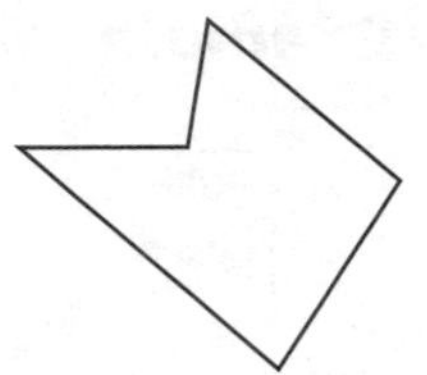

D. 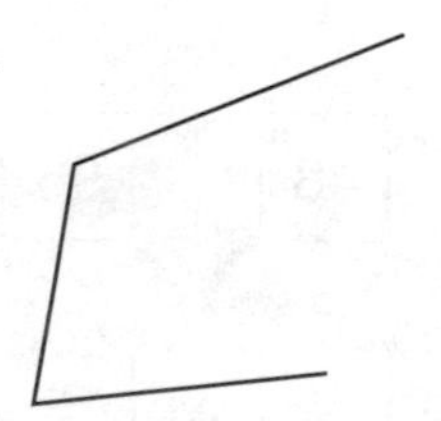

E. 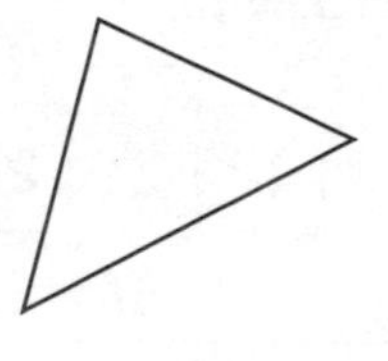

F. 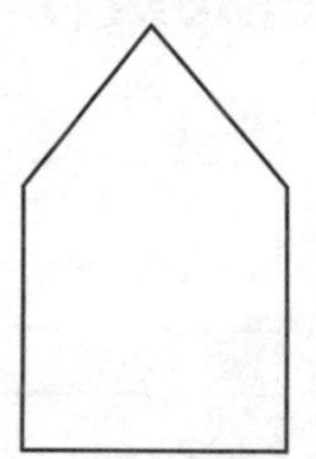

Write 2 addition and 2 subtraction facts for the following groups of numbers.

2. 2, 9, 11

3. 8, 9, 17

4. 1, 6, 7

5. 3, 4, 7

6. 4, 6, 10

7. 6, 7, 13

Write the number models with parentheses and solve.

8. Add 15 to the difference of 105 and 70.

9. Subtract the sum of 8 and 3 from 18.

10. Add 9 to the difference of 50 and 16.

11. Subtract the sum of 81 and 42 from 338.

Practice Set 4 (cont.)

For each Fact Minute below, do as many problems as you can in that minute. You can ask someone to time you.

Fact Minute 1

12. $18 \div 6$

13. 7×7

14. 4×5

15. $54 \div 9$

16. 3×4

17. $64 \div 8$

18. 6×8

19. 6×6

20. 9×9

21. $49 \div 7$

22. 9×4

23. 7×9

24. 6×2

25. $56 \div 7$

26. 7×6

Fact Minute 2

27. 4×8

28. $36 \div 6$

29. 5×9

30. 8×4

31. 3×8

32. $48 \div 6$

33. 7×4

34. $28 \div 7$

35. 2×5

36. 6×3

37. $32 \div 4$

38. 6×7

39. 4×7

40. $20 \div 5$

41. 8×9

Fact Minute 3

42. 9×6

43. $24 \div 4$

44. 2×7

45. 8×7

46. 3×7

47. $27 \div 3$

48. 7×5

49. 3×6

50. $8 \div 4$

51. 4×3

52. 9×8

53. $63 \div 9$

54. 9×5

55. $15 \div 3$

56. 2×8

Practice Set 5

Write all of your answers on a separate sheet of paper.

1. Which figures are regular polygons?

A. 

B. 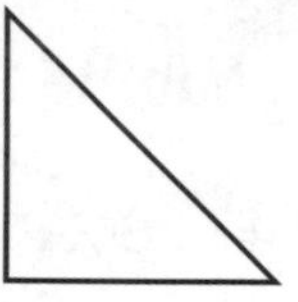

C. 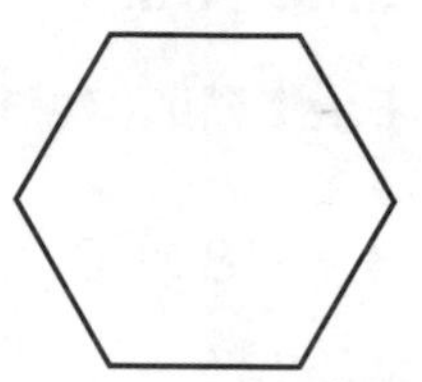

D. 

E. 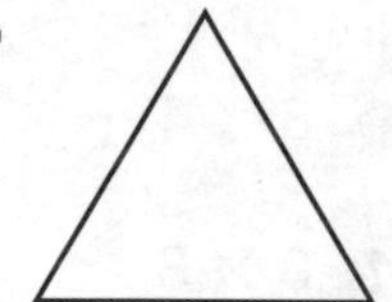

F. 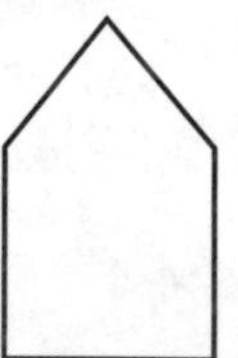

Solve.

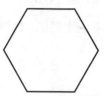

2. $120 - \blacksquare = 30$

3. $82 + 17$

4. $70 + 26$

5. $40 - 17$

6. $16 - 8$

7. 9×9

8. $\begin{array}{r} 87 \\ -\ 36 \\ \hline \end{array}$

9. $\begin{array}{r} 33 \\ -\ 14 \\ \hline \end{array}$

10. $\begin{array}{r} 37 \\ \times\ 8 \\ \hline \end{array}$

11. $\begin{array}{r} 11 \\ +\ 34 \\ \hline \end{array}$

12. $\begin{array}{r} 120 \\ \times\ 50 \\ \hline \end{array}$

13. $\begin{array}{r} 521 \\ +\ 131 \\ \hline \end{array}$

14. $\begin{array}{r} 34 \\ +\ 15 \\ \hline \end{array}$

15. $\begin{array}{r} 35 \\ \times\ 3 \\ \hline \end{array}$

Practice Set 6

Write all of your answers on a separate sheet of paper.

Match each description with the correct figure. Write the letter that identifies the figure.

1. concentric circles

2. right triangle

3. rhombus

4. rectangle

A.

B.

C.

D.

Solve.

5. Joy wants to have enough balloons for her 22 party guests. How many packages of 6 does she need?

6. Larry has a large pizza to share with 3 friends. If the pizza is divided into 16 slices, how many slices will each person, including Larry, get?

7. If four more friends join Larry and the others, how many slices will each person get?

Complete the frames-and-arrows problem.

8.

Rule
× 2

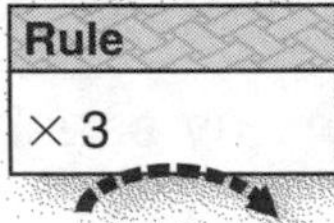

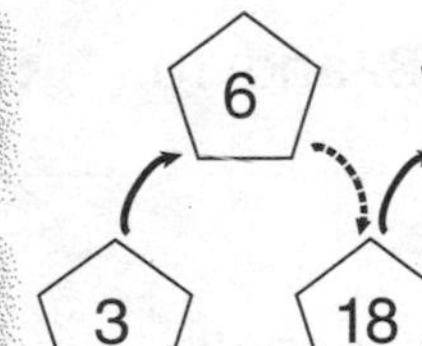

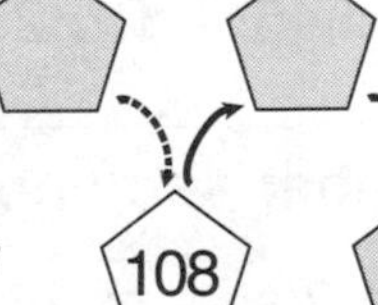

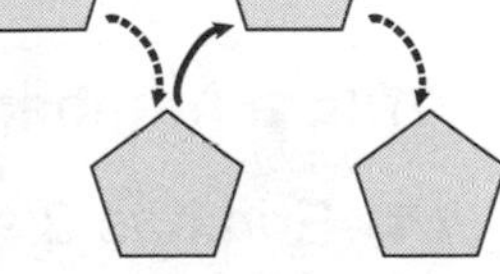

Practice Set 7

Write all of your answers on a separate sheet of paper.

Find the area, in square units, of each rectangle, and then write the number model.

Reminder: Area = length (l) × width (w)

Example

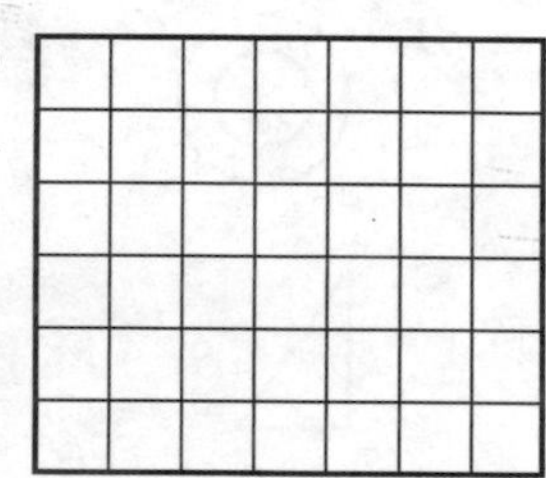

$6 \times 7 = 42$

1.

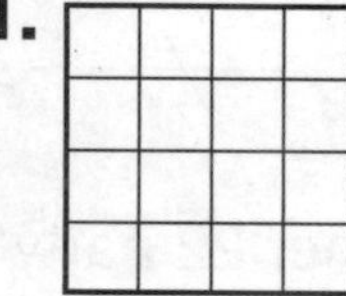

2.

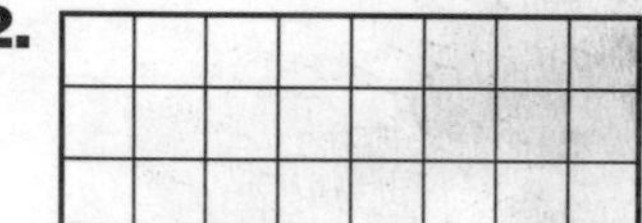

Write the digit in the hundredths place for each of the following.

3. 5.925 **4.** 1.043 **5.** 8.100 **6.** 0.280 **7.** 3.313

Write 2 multiplication and 2 division facts for each group of numbers.

8. 6, 7, and 42 **9.** 3, 9, and 27 **10.** 4, 8, and 32

11. 5, 9, and 45 **12.** 2, 8, and 16 **13.** 4, 7, and 28

Write a number sentence.

14. Erin has 3 sets of 25 seashells. How many seashells does Erin have in all?

Practice Set 8

Write all of your answers on a separate sheet of paper.

Make a name collection box for each number listed below. Use as many different numbers as you can.

Example

42
50 − 8
20 × 2 + 2
100 ÷ 4 + 17
3 × 2 × 7
$\frac{84}{2}$

1. 27

2. 125

3. 64

4. 180

5. 32

6. 85

Solve.

7. 6 × 6

8. 12 × 6

9. 7 × 5

10. 10 × 7

11. 11 × 4

12. 9 × 9

13. 6 × 8

14. 7 × 5

15. 9 × 11

16. 2 × 12

17. 11 × 8

18. 8 × 12

19. 10 × 11

20. 8 × 4

21. 60 ÷ 10

22. 9 × 5

23. 7 × 11

24. 12 × 7

25. 16 ÷ 8

26. 60 ÷ 12

Practice Set 9

Write all of your answers on a separate sheet of paper.

Add.

1.
$$\begin{array}{r} 700 \\ 40 \\ + \quad 2 \\ \hline \end{array}$$

2.
$$\begin{array}{r} 8{,}000 \\ 200 \\ 30 \\ + \quad 1 \\ \hline \end{array}$$

3.
$$\begin{array}{r} 60{,}000 \\ 500 \\ 70 \\ + \quad 3 \\ \hline \end{array}$$

4.
$$\begin{array}{r} 90{,}000 \\ 6{,}000 \\ 800 \\ 10 \\ + \quad 4 \\ \hline \end{array}$$

5.
$$\begin{array}{r} 800{,}000 \\ 3{,}000 \\ 300 \\ + \quad 1 \\ \hline \end{array}$$

6.
$$\begin{array}{r} 4{,}000{,}000 \\ 900{,}000 \\ 30{,}000 \\ 400 \\ + \quad 90 \\ \hline \end{array}$$

---

7. Use the clues to complete the place-value puzzle.

- Divide 72 by 6. Subtract 4 and write the result in the ones place.
- Double the number in the ones place and divide by 8. Write the result in the tens place.
- Multiply 9×10. Subtract 83. Write the result in the hundreds place.
- Halve the number in the tens place. Multiply by 3 and write the result in the thousands place.
- Divide 27 by the number in the thousands place. Write the result in the ten-thousands place.

10,000s	1,000s	100s	10s	1s

Practice Set 9 (cont.)

Write all of your answers on a separate sheet of paper.

In each set of problems below, do as many exercises as you can in one minute. Ask someone to time you.

Problem Set 1

8. 10×6

9. 7×11

10. $48 \div 4$

11. 2×9

12. 10×4

13. $90 \div 9$

14. 5×11

15. 10×12

16. $49 \div 7$

17. 4×12

18. 7×8

19. $63 \div 9$

20. $45 \div 5$

21. 6×7

22. $18 \div 2$

Problem Set 2

23. 2×5

24. 9×11

25. $110 \div 10$

26. 8×4

27. 11×10

28. 6×12

29. 11×11

30. 12×11

31. 4×7

32. 12×16

33. $63 \div 7$

34. 8×6

35. 3×3

36. $45 \div 9$

37. $16 \div 4$

Problem Set 3

38. $54 \div 9$

39. 12×2

40. 12×5

41. 12×3

42. $121 \div 11$

43. 60×5

44. 8×7

45. $144 \div 12$

46. $8 \div 4$

47. $33 \div 3$

48. $42 \div 7$

49. $81 \div 9$

50. $32 \div 8$

51. $18 \div 9$

52. 7×3

Practice Set 10

Write all of your answers on a separate sheet of paper.

Use digits to write the following numbers.

1. twenty-four thousand, nine hundred sixty-eight

2. seventy-six thousand, six hundred fourteen

3. six thousand, nine hundred two

Write the words for the following numbers.

4. 12,743

5. 8,054

6. 69,231

7. 4,782

Solve.

8. $\begin{array}{r} 26 \\ +\ 47 \\ \hline \end{array}$

9. $\begin{array}{r} 63 \\ +\ 18 \\ \hline \end{array}$

10. $\begin{array}{r} 16 \\ \times\ 4 \\ \hline \end{array}$

11. $\begin{array}{r} 180 \\ \times\ 7 \\ \hline \end{array}$

12. $\begin{array}{r} 196 \\ \times\ 0 \\ \hline \end{array}$

13. $\begin{array}{r} 32.1 \\ +\ 18.7 \\ \hline \end{array}$

14. $\begin{array}{r} 1.25 \\ +\ 6.43 \\ \hline \end{array}$

15. $\begin{array}{r} 8.40 \\ -\ 5.01 \\ \hline \end{array}$

16. $\begin{array}{r} 11 \\ \times\ 9 \\ \hline \end{array}$

17. $\begin{array}{r} 85 \\ -\ 38 \\ \hline \end{array}$

18. $\begin{array}{r} 20 \\ 83 \\ +\ 17 \\ \hline \end{array}$

19. $\begin{array}{r} 83 \\ -\ 41 \\ \hline \end{array}$

Practice Set 11

Write all of your answers on a separate sheet of paper.

The tally chart at the right shows the number of items that some fourth graders missed on a quiz.

Number of Items Missed	Number of Students
0	~~////~~ //
1	~~////~~ /
2	///
3	//
4	//
5	/
6	//
7	/

1. How many students reported the number of items they missed?

2. What is the *maximum* (largest) number of items missed?

3. What is the *minimum* (smallest) number of items missed?

4. What is the *range*?

5. What is the *mode* (most frequent) number of items missed?

Use digits to write the following numbers:

6. sixteen thousand, five hundred forty-seven

7. eight and two-tenths

8. seven and nine-tenths

Write the words for the following numbers:

9. 21,894

10. 14.1

11. 48,563

12. 903

Practice Set 11 (cont.)

Write all of your answers on a separate sheet of paper.

Complete the "What's My Rule?" tables.

13. Rule: out = in * 20

in	out
9	180
12	240
15	
25	
100	

14. Rule:

in	out
7	3.5
10	6.5
	10.5
16.5	
20.5	17

15. Rule:

in	out
80	20
160	
	90
2,400	
4,800	1,200

16. Rule: out = in * 10

in	out
3	
6	
	90
	120
15	

Rewrite the number sentences with parentheses to make them correct.

17. 6 * 11 − 7 = 59

18. 2.2 = 8 − 3 + 2.8

19. 330 − 150 − 60 = 240

20. 18 = 2 * 5.4 + 3.6

21. 7 * 2.1 + 5 * 12 = 74.7

22. 230 = 4 * 60 − 10

23. 3 * 9 + 3 − 4 = 32

24. 584 = 11 * 50 + 34

Practice Set 12

Write all of your answers on a separate sheet of paper.

Mr. Adema asked his piano students to estimate the number of hours they practice each week. The tally chart shows the data he collected. Use the table to help you answer the questions below.

Number of Hours	Number of Students
2	//
3	~~////~~ //
4	////
5	///
6	//
7	/
8	/

1. Construct a line plot for the data.
2. What is the maximum number of hours spent practicing each week?
3. What is the minimum number of hours spent practicing each week?
4. What is the range?
5. What is the median number of hours?

Complete the frames-and-arrows problems.

6. 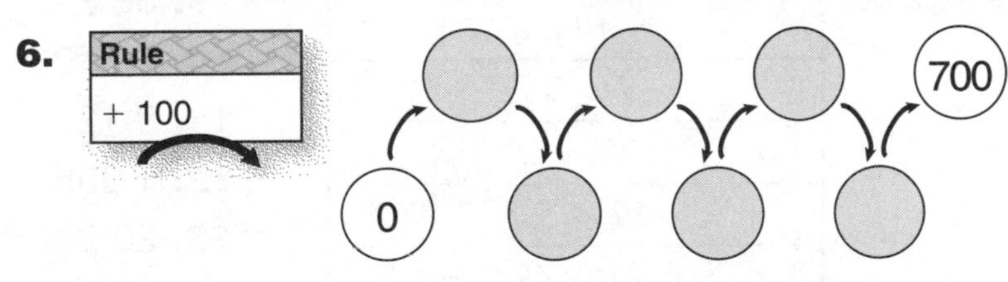

7. 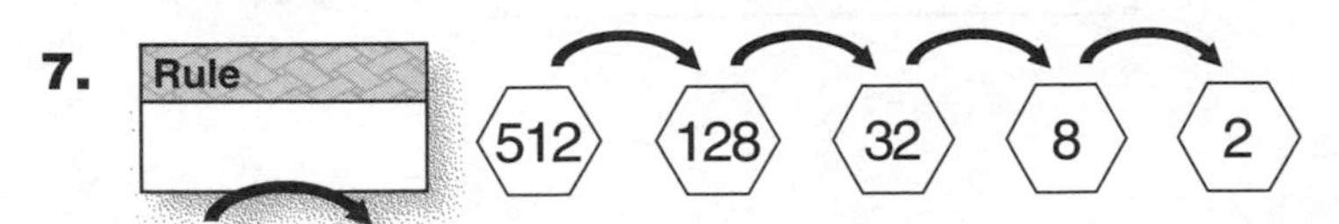

Write all of your answers on a separate sheet of paper.

8. Use the clues to complete the place-value puzzle.

- Divide 88 by 11. Add 1 and write the result in the thousands place.
- Double the number in the thousands place and divide by 3. Write the result in the tens place.
- Multiply 4 * 12. Subtract 42. Write the result in the hundreds place.
- Divide 63 by the number in the thousands place. Write the result in the ones place.
- Halve the number in the tens place. Add 1 and write the result in the ten-thousands place.

10,000s	1,000s	100s	10s	1s

Make a name-collection box for each number below. Use as many different numbers and operations as you can.

Example

14.2
71 ÷ 5
*7.1 * 2*
20 − 5.8
*(3.5 * 2) + (9.2 − 2)*

9. 38.7

10. 7,049

11. 8.12

12. 1,005

13. 20.2

14. 706

Practice Set 13

Write all of your answers on a separate sheet of paper.

Solve. Use the partial sums method. Show your work.

1. 204 + 149

2. 551 + 267

3. 859 + 1,596

4. 1,886 + 4,269

Solve. Use the column addition method. Show your work.

5. 734 + 478

6. 592 + 879

7. 2,735 + 1,305

8. 6,371 + 19,583

Solve. Use any method you choose.

9. 795 + 616

10. 8,214 + 5,488

11. 5,838 + 8,956

12. 50,694 + 39,518

Complete the missing factors.

13. 7 * ■ = 21

14. ■ * 4 = 36

15. ■ * 8 = 64

16. 12 * ■ = 96

17. 400 * ■ = 3,600

18. ■ * 5 = 350

19. 9 * ■ = 810

20. ■ * 6 = 660

Estimate the total cost.

21. 12 rulers that cost $1.05 each

22. 4 scissors that cost $0.69 each

23. 7 books that cost $3.45 each

Practice Set 14

Write all of your answers on a separate sheet of paper.

Mrs. Lewis teaches art. She made a graph to show the number of art projects the students have completed. Use the bar graph to find the following landmarks for the data.

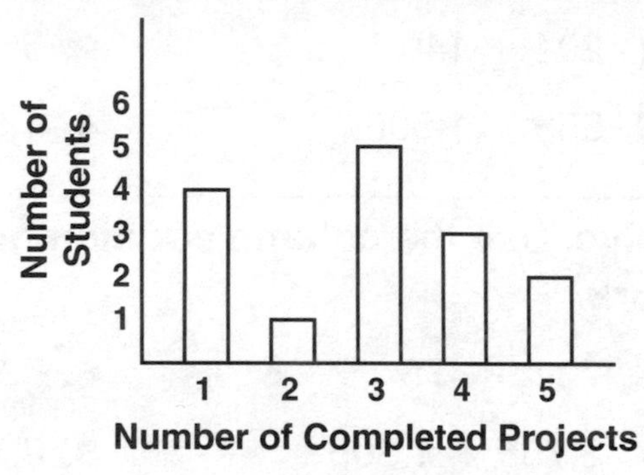

1. What is the maximum number of completed projects?

2. What is the minimum number of completed projects?

3. What is the range?

4. What is the median?

Measure the line segments to the nearest cm.

5. ______________________________

6. ________________________

7. ____________________________

8. ____________

Practice Set 14 (cont.)

Write all of your answers on a separate sheet of paper.

Complete the frames-and-arrows problems.

9.

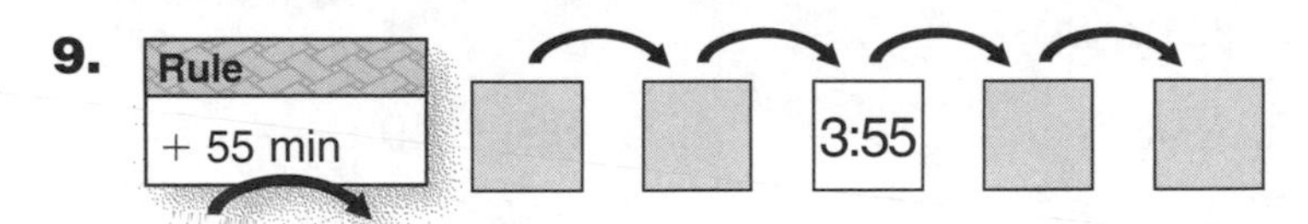

10.

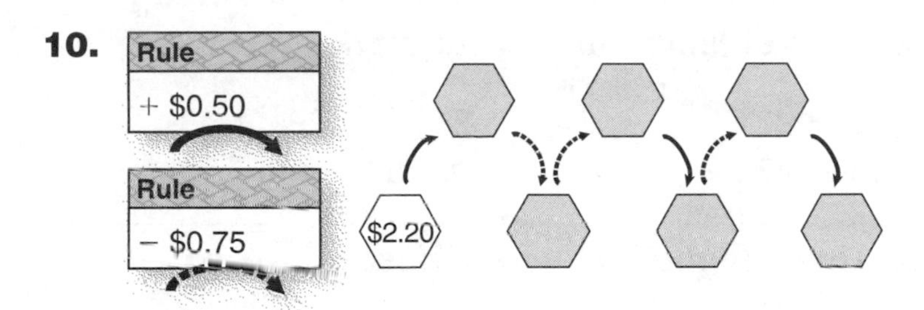

11.

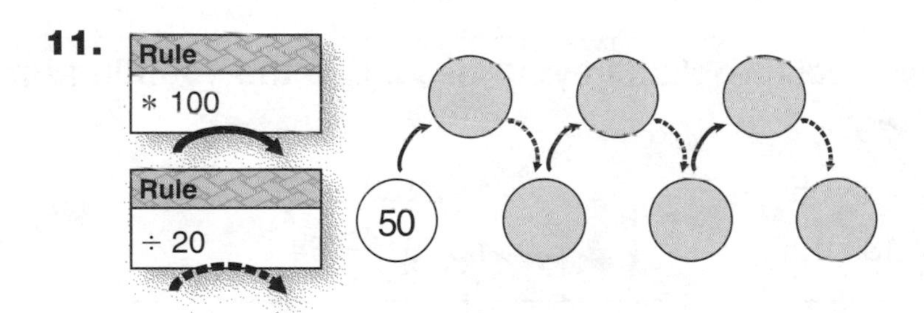

12.

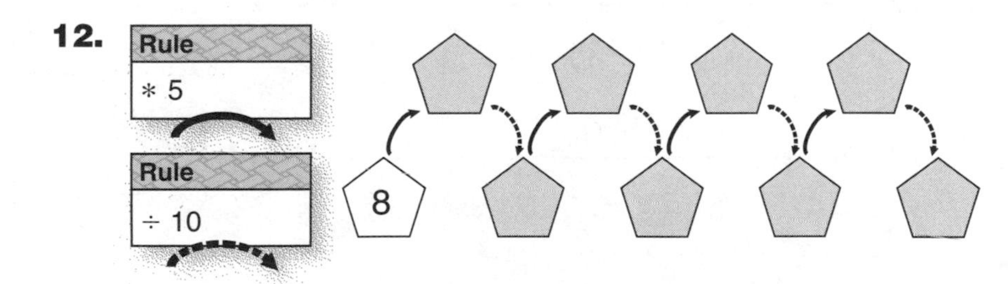

Practice Set 15

Write all of your answers on a separate sheet of paper.

Use the trade-first method to solve the problems. Show your work.

1. 92 – 37 **2.** 85 – 49

3. 624 – 286 **4.** 348 – 159

Use the partial differences subtraction method to solve the problems. Show your work.

5. 53 – 29 **6.** 77 – 58

7. 658 – 270 **8.** 734 – 386

Solve. Use any method you choose.

9. 79 – 23 **10.** 33 – 17

11. 636 – 498 **12.** 961 – 185

Solve each problem in your head. Use the counting-up strategy.

13. 70 – 51 **14.** 130 – 97

15. 48 – 20 **16.** 91 – 54

Fill in the missing numbers on the number lines.

17.
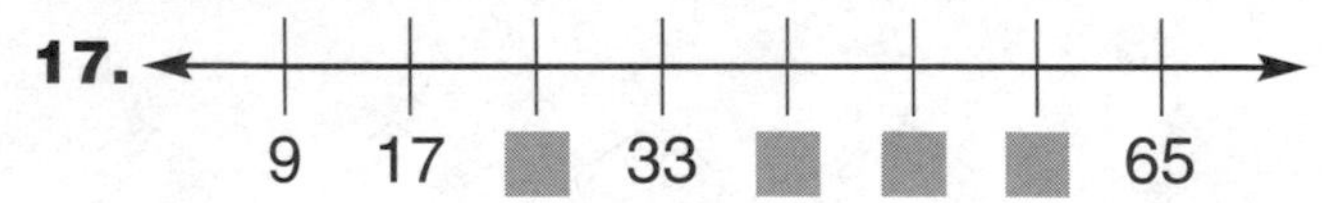

18.
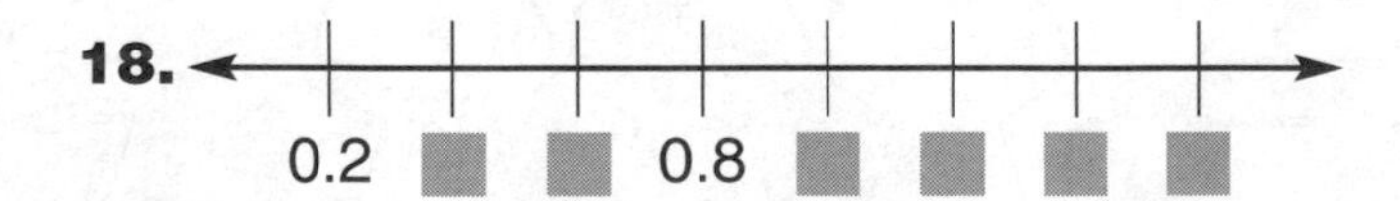

19.
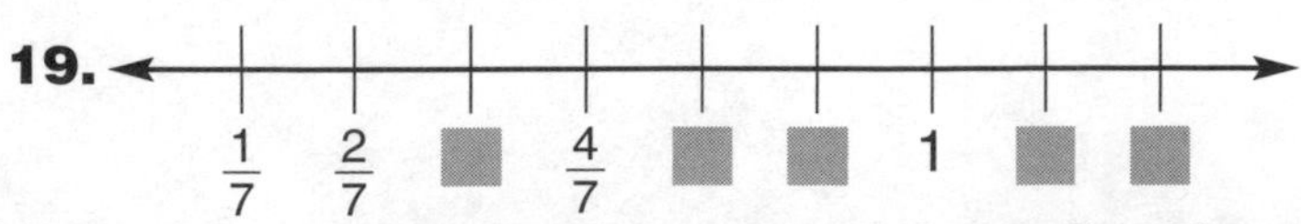

Practice Set 15 (cont.)

Write all of your answers on a separate sheet of paper.

20. How many pieces of fruit are there?

21. What fraction of the fruit is apples?

22. What fraction of the fruit is pears?

23. What fraction of the fruit is bananas?

Solve.

24. $67 * 4$

25. $53 * 8$

26. $84 \div 3$

27. $675 * 6$

28. $8{,}229 + 3{,}160$

29. $7{,}583 - 5{,}432$

30. $3{,}499 * 0$

31. $673 - 64$

32. $467 + 185$

33. $320 * 8$

34. $560 \div 8$

35. $8{,}524 - 1{,}996$

Practice Set 16

Write all of your answers on a separate sheet of paper.

In each set of problems below, do as many exercises as you can in one minute. Ask someone to time you.

Problem Set 1

1. 9 * 6

2. 7 * 7

3. 21 ÷ 7

4. 12 * 8

5. 44 ÷ 4

6. 2 * 10

7. 11 * 4

8. 64 ÷ 8

9. 12 * 5

10. 10 * 11

11. 81 ÷ 9

12. 54 ÷ 6

13. 9 * 7

14. 48 ÷ 8

15. 9 * 3

Problem Set 2

16. 12 * 7

17. 2 * 9

18. 6 * 6

19. 9 * 9

20. 121 ÷ 11

21. 6 * 7

22. 4 * 12

23. 21 ÷ 3

24. 108 ÷ 9

25. 8 * 4

26. 42 ÷ 6

27. 144 ÷ 12

28. 4 * 10

29. 11 * 11

30. 8 * 3

Problem Set 3

31. 12 * 11

32. 54 ÷ 9

33. 42 ÷ 7

34. 7 * 3

35. 12 * 4

36. 55 ÷ 5

37. 6 * 8

38. 3 * 11

39. 90 ÷ 9

40. 48 ÷ 6

41. 12 * 9

42. 4 * 7

43. 3 * 8

44. 132 ÷ 12

45. 49 ÷ 7

Write all of your answers on a separate sheet of paper.

Lightbulbs
4-pack $1.07

Tissues
$0.99

Batteries
4-pack $1.99

Transparent Tape
$0.84

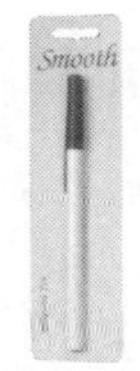

Ballpoint Pen
$0.24

VCR Tape
$2.79

46. John must buy supplies for his company. He needs five lightbulbs, four boxes of tissues, and six rolls of transparent tape. How much money will he need for these supplies?

47. Ms. Larson has four dollars. How many pens can she buy?

48. Judy and Sarah are going to videotape their school's pageant. They need eight batteries and two VCR tapes for the camera. They have ten dollars. Do they have enough money to buy what they need? What is the difference between the money they have and the money they need?

49. About how much is each of the lightbulbs in the 4-pack?

50. About how much is each of the batteries in the 4-pack?

Practice Set 17

Write the missing numbers on a separate sheet of paper.

1. 3 * ? = 15

2. ? * 8 = 48

3. ? / 7 = 6

4. 3 * ? = 30

5. 18 / ? = 9

6. ? * 7 = 63

7. ? * 4 = 32

8. ? / 2 = 7

9. 20 / ? = 5

10. ? * 5 = 25

11. ? / 7 = 4

12. 54 / ? = 6

Who am I?

13. Clue 1: I am less than 10.

Clue 2: I am an odd number.

Clue 3: If you turn me upside down, I am an even number.

14. Clue 1: I am less than 100.

Clue 2: The sum of my digits is 8.

Clue 3: If you divide me by 2, I am an even number.

Clue 4: My tens digit and my ones digit are the same.

15. Clue 1: I am a number between 75 and 150.

Clue 2: My tens digit is three times my ones digit.

Clue 3: The sum of my digits is 5.

Clue 4: My hundreds digit and my ones digit are the same.

Practice Set 18

Write all of your answers on a separate sheet of paper.

Find the missing number for each Fact Triangle. Then write the fact family for that triangle.

Example

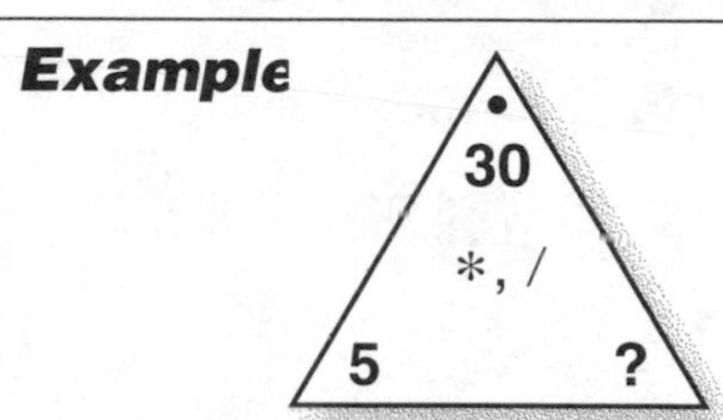

Missing number: 6

Fact family: $5 * 6 = 30$
$6 * 5 = 30$
$30 / 6 = 5$
$30 / 5 = 6$

1.

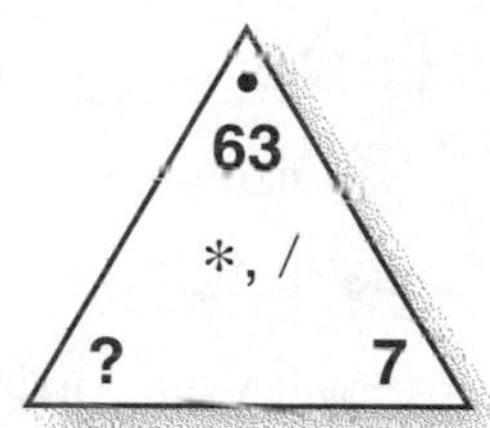

2.

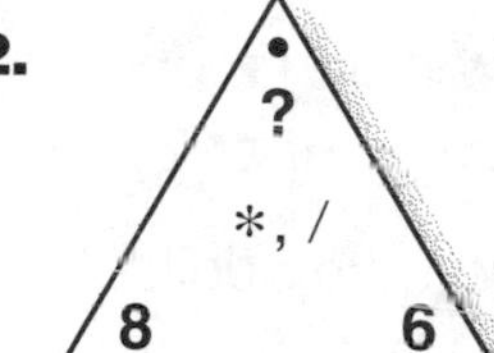

3.

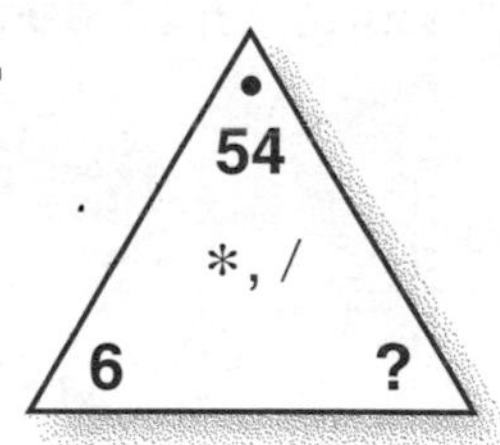

4.

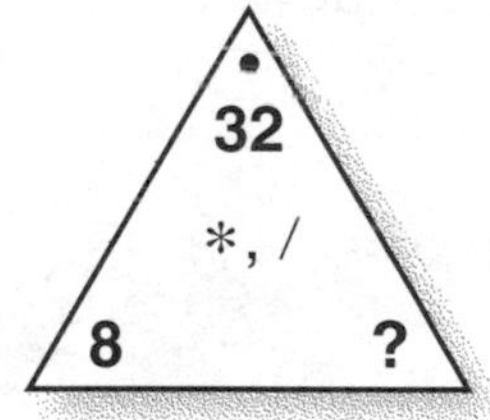

5.

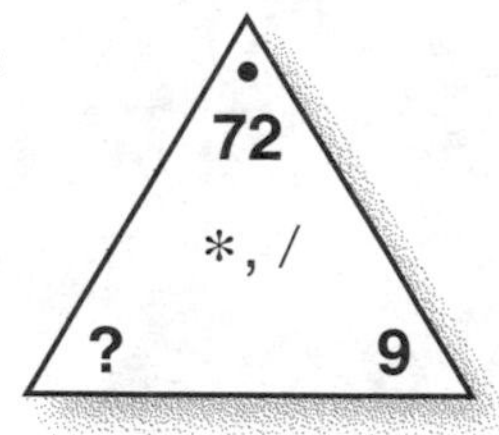

6.

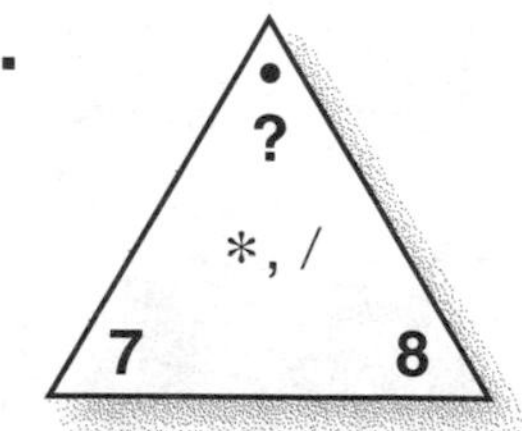

Practice Set 18 (cont.)

Write all of your answers on a separate sheet of paper.

Solve.

7. $80 \div 8 = \blacksquare$

8. $30 * 80 = \blacksquare$

9. $800 = 8 * \blacksquare$

10. $4 * 400 = \blacksquare$

11. $30 * \blacksquare = 1{,}500$

12. $\blacksquare \div 1{,}000 = 8$

13. $1{,}400 \div 700 = \blacksquare$

14. $28 * \blacksquare = 560$

15. $\blacksquare \div 70 = 70$

16. $6 * 30 = \blacksquare$

17. $4{,}500 \div \blacksquare = 5$

18. $9 * 90 = \blacksquare$

19. How much money, without tax, will I need for 3 boxes of crackers that cost $1.59 each?

20. How many dollars in 18 five-dollar bills?

21. If 1 block is 200 meters long, how far will you run in 7 blocks?

Write numbers for the fractional parts shown in each picture.

Example

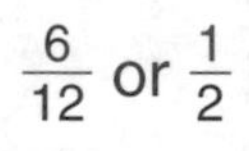

22.

23.

24.

25.

Practice Set 19

Write all of your answers on a separate sheet of paper.

Solve. Write a number model.

1. In May Mr. Tong drove his car 1,714 miles. In June, he drove 946 miles. How many miles did he drive in all during those two months?

2. The T-shirt Mart sells small, medium, and large T-shirts. There are 342 small T-shirts, 496 medium T-shirts, and 683 large T-shirts in stock. How many more large T-shirts are in stock than small T-shirts?

3. Ellen has two pieces of string. One is 143 cm in length. The other is 257 cm in length. What is the difference between the lengths of the two pieces?

Solve.

4. $\begin{array}{r} 440 \\ 115 \\ +\ 711 \\ \hline \end{array}$

5. $\begin{array}{r} 79 \\ +\ 28 \\ \hline \end{array}$

6. $\begin{array}{r} 784 \\ -\ 426 \\ \hline \end{array}$

7. $\begin{array}{r} 230 \\ *\ 8 \\ \hline \end{array}$

8. $\begin{array}{r} 112 \\ *\ 9 \\ \hline \end{array}$

9. $\begin{array}{r} 263 \\ 357 \\ +\ 198 \\ \hline \end{array}$

10. $\begin{array}{r} 4{,}315 \\ -\ 78 \\ \hline \end{array}$

11. $\begin{array}{r} 24 \\ -\ 11 \\ \hline \end{array}$

12. $\begin{array}{r} 78 \\ *\ 4 \\ \hline \end{array}$

13. $\begin{array}{r} 625 \\ -\ 36 \\ \hline \end{array}$

14. $\begin{array}{r} 482 \\ *\ 2 \\ \hline \end{array}$

15. $\begin{array}{r} 96 \\ *\ 3 \\ \hline \end{array}$

Write all of your answers on a separate sheet of paper.

Write the amounts.

16. (Q) (Q) (Q) (Q) (Q) (D) (D) (N) (N) (N) (P) (P) (P)

17. [$1] [$1] (Q) (Q) (Q) (D) (D) (D) (D) (N) (P) (P)

18. [$5] [$5] [$5] [$1] (Q) (N) (N) (N)

19. [$100] [$20] [$20] [$5] [$1] [$1] [$1] (Q)

Solve.

20. Mrs. Brown's class kept track of the number of hours they spent reading each day. The graph shows the number of hours the students spent reading Monday through Wednesday.

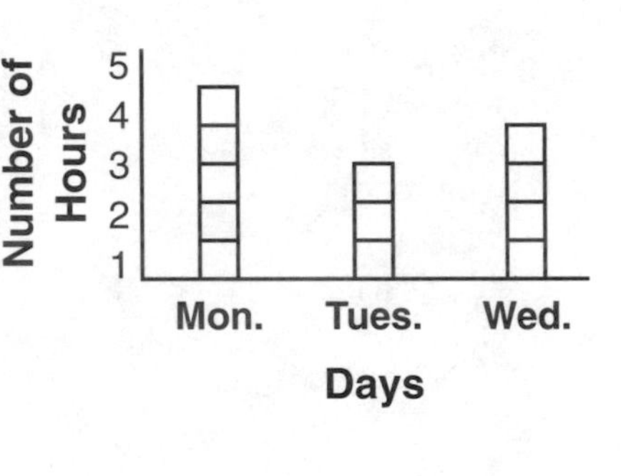

a. How many more hours did they read on Monday than on Tuesday?

b. What is the average number of hours they spent reading in a day?

c. How many total hours do you think they would read, Monday through Friday? Explain.

Practice Set 20

Write all of your answers on a separate sheet of paper.

For each number sentence, write *T* if it is true, *F* if it is false, or *?* if you can't tell.

1. $8 * 9 = 76$

2. $5 + 9 < 20$

3. $4 = 64 / 8$

4. $26 + 19 = 7$

5. $450 - 119$

6. $18 < 15 + 6$

7. $70 - 21 = 49$

8. $9 * 4 > 36$

Write a number for each picture below. Use 0 or $\frac{0}{4}$, $\frac{1}{4}$, $\frac{1}{2}$ or $\frac{2}{4}$, $\frac{3}{4}$, and 1 or $\frac{4}{4}$.

9.

10.

11.

12.

13.

Solve.

14. $\begin{array}{r} 47 \\ * \ 6 \\ \hline \end{array}$

15. $\begin{array}{r} 63 \\ * \ 3 \\ \hline \end{array}$

16. $\begin{array}{r} 214 \\ * \ 5 \\ \hline \end{array}$

17. $\begin{array}{r} 703 \\ * \ 7 \\ \hline \end{array}$

Practice Set 21

Write all of your answers on a separate sheet of paper.

Rewrite the number models with parentheses to make them correct.

1. 6 * 8 − 3 = 45

2. 22 = 8 + 3 * 2

3. 33 − 15 − 6 = 24

4. 54 − 10 + 8 = 52

5. 3 * 8 + 2 * 11 = 46

6. 30 = 4 * 6 + 6

7. 2 * 2 + 7 * 8 = 60

8. 489 = 5 * 25 + 75 − 11

---

9. In baseball, the bases on the diamond are placed exactly 90 ft apart.

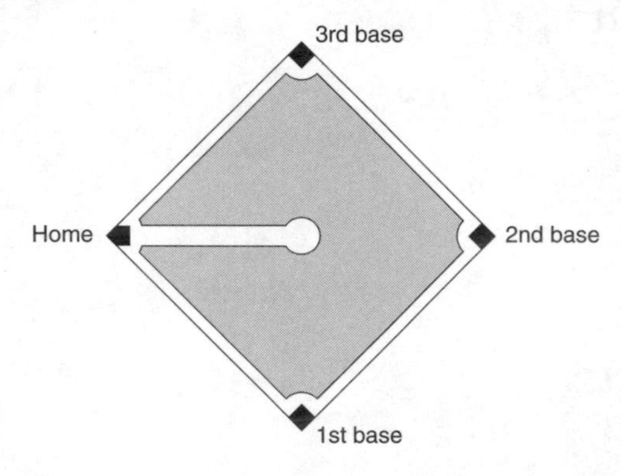

a. If a batter hits a home run, how many feet does she run?

b. If there are runners on first and third when a batter hits a home run, what is the total distance all three players run?

10. Write the following number in digits: eight thousand, four hundred twenty-one.

11. Write the words for 1,603.

Practice Set 22

Write all of your answers on a separate sheet of paper.

Find the solution of each open sentence below. Write a number sentence with the solution in place of the variable.

1. $x + 8 = 35$

2. $4t = 24$

3. $140 + 3 = y$

4. $s / 7 = 7$

5. $m - 60 = 200$

6. $68 + r = 80$

7. $6x = 42$

8. $70 / n = 10$

Rewrite the number sentences with parentheses to make them correct.

9. $204 = 7 * 20 + 75 - 11$

10. $7 * 9 - 4 = 35$

11. $42 = 3 + 3 * 7$

12. $31 - 15 - 6 = 10$

13. $54 - 10 + 8 = 52$

14. $7 * 8 + 3 * 11 = 89$

Solve.

15. $\begin{array}{r} 212 \\ * \ 20 \\ \hline \end{array}$

16. $\begin{array}{r} 785 \\ - \ 76 \\ \hline \end{array}$

17. $\begin{array}{r} 867 \\ - \ 74 \\ \hline \end{array}$

18. $\begin{array}{r} 900 \\ + \ 1{,}200 \\ \hline \end{array}$

19. $\begin{array}{r} 418 \\ 460 \\ + \ 454 \\ \hline \end{array}$

20. $\begin{array}{r} 1{,}034 \\ + \ 2{,}349 \\ \hline \end{array}$

21. $\begin{array}{r} 76 \\ * \ 0 \\ \hline \end{array}$

22. $\begin{array}{r} 7{,}210 \\ + \ 9{,}188 \\ \hline \end{array}$

23. $\begin{array}{r} 600 \\ - \ 599 \\ \hline \end{array}$

24. $(60 + 80) * 4$

25. $39 - (3 * 4)$

26. $620 + 150 + 220$

27. $(1{,}800 \div 60) * 5$

Practice Set 22 (cont.)

Write all of your answers on a separate sheet of paper.

28. The first figure is $\frac{1}{2}$ of the whole. What fraction of the whole is each of the other figures?

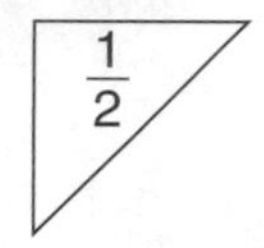

a. 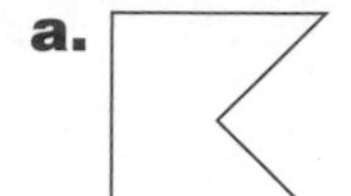

b. 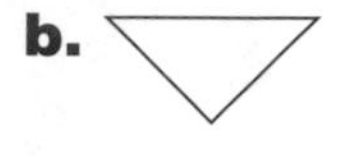

c. 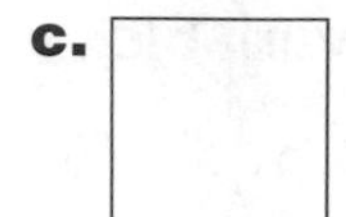

The fourth graders had a pizza party. They ordered pizzas and divided each pizza into 6 equal slices. Twenty-one students, 1 teacher, and 4 parents were invited to the party. The pupils assumed each person would eat one slice of pizza.

29. How many people were invited to the party?

30. How many slices of pizza did they need?

31. How many pizzas did the class order?

32. If everyone ate just one slice, how many slices were left over?

33. What fraction of a whole pizza is that?

34. If everyone ate two slices of pizza, how many slices did they need?

35. How many whole pizzas did the class then need?

36. What fraction of a whole pizza was left over?

37. Juana brought 3 granola bars to divide equally among 4 of her friends and herself. What fraction of one granola bar did each person get?

Practice Set 23

Write all of your answers on a separate sheet of paper.

Write the digit in the hundredths place.

1. 5.92 **2.** 1.04 **3.** 1.10

4. 8.10 **5.** 0.280 **6.** 3.02

7. 3.31 **8.** 0.01 **9.** 10.12

Write the numbers in order from smallest to largest.

10. 1.2, 0.2, 2.10, 2.2

11. 0.23, 1.2, 0.04, 5.1

12. 4.01, 1.4, 2.14, 1.41

13. 9.5, 1.95, 19.5, 0.59

14. 0.3, 3.3, 0.03, 3.03

15. 5.20, 5.12, 5.02, 5.21

Solve.

16. 420 − ■ = 81

17. 712 + 517 = ■

18. 160 + 348 = ■

19. 490 − 170 = ■

20. 2,216 − 1,804 = ■

21. 90 = 8,100 ÷ ■

22. $\begin{array}{r} 387 \\ -\ 36 \\ \hline \end{array}$

23. $\begin{array}{r} 673 \\ -\ 615 \\ \hline \end{array}$

24. $\begin{array}{r} 57 \\ *\ 7 \\ \hline \end{array}$

25. $\begin{array}{r} 7{,}619 \\ +\ 3{,}250 \\ \hline \end{array}$

26. $\begin{array}{r} 980 \\ *\ 50 \\ \hline \end{array}$

27. $\begin{array}{r} 427 \\ 561 \\ +\ 711 \\ \hline \end{array}$

28. Alvin wants to wear a different pair of socks for each of the 14 days he will be on vacation. How many socks does he need to pack?

Practice Set 23 (cont.)

Write all of your answers on a separate sheet of paper.

Complete the frames-and-arrows problems.

29.

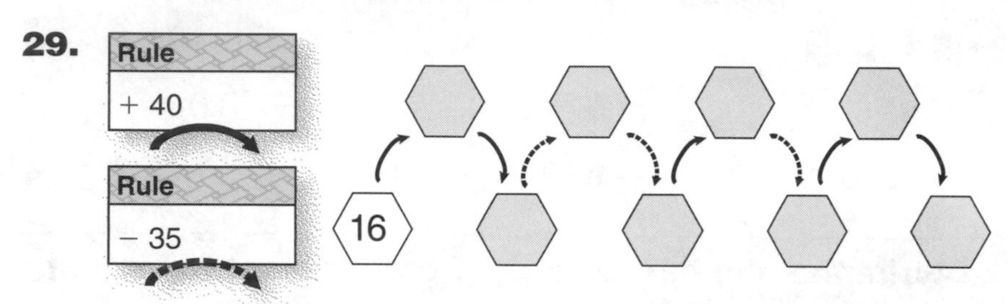

30.

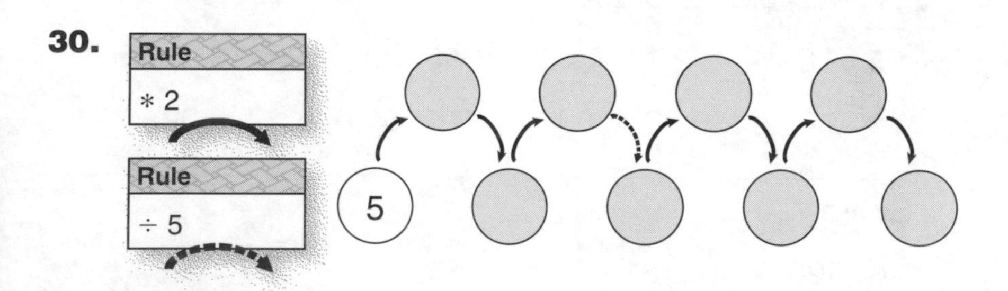

31.

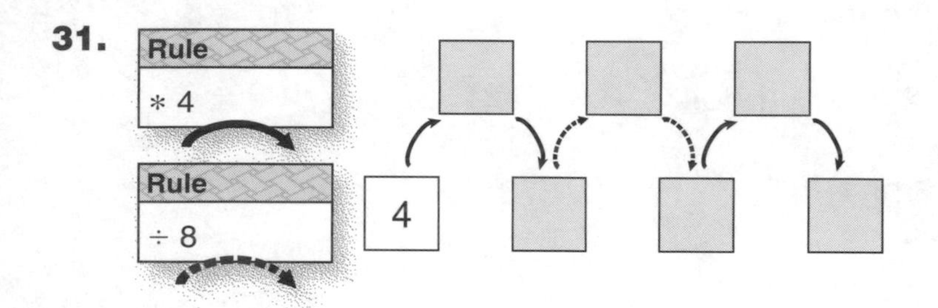

32. Write the following in digits: one hundred sixty-two thousand, nine hundred seventy-four.

33. Write the words for 171,603.

34. Write the following in digits: two hundred thousand, eight hundred forty-four.

Practice Set 24

Write all of your answers on a separate sheet of paper.

Mrs. Gomez is a food service director. She bought fruits and vegetables shown in the table for a PTA dinner. Use estimation to answer the following questions.

Item	Amount
apples	15.2 kg
oranges	19.7 kg
watermelon	17.3 kg
celery	12.1 kg
potatoes	33.8 kg
tomatoes	6.9 kg

1. Which item did she buy the most of?

2. Which item did she buy the least of?

3. About how many more kilograms of oranges did she buy than kilograms of apples?

---

4. Use the clues to complete the place-value puzzle.

- Write the result of 210 ÷ 70 in the ten-thousands place.
- Multiply 5 * 7. Subtract 29. Write the result in the tens place.
- Triple the number in the ten-thousands place. Write the result in the thousands place.
- Divide 72 by 12. Add 1 and write the result in the hundreds place.
- Subtract the number in the ten-thousands place from the number in the hundreds place. Write the result in the hundred-thousands place.
- Multiply 84 by 0. Write the result in the ones place.

100,000s	10,000s	1,000s	100s	10s	1s

Practice Set 24 (cont.)

Complete the frames-and-arrows problems.

5.

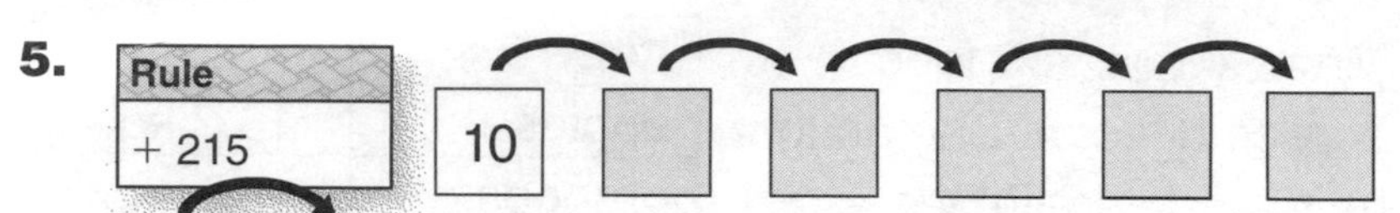

6.

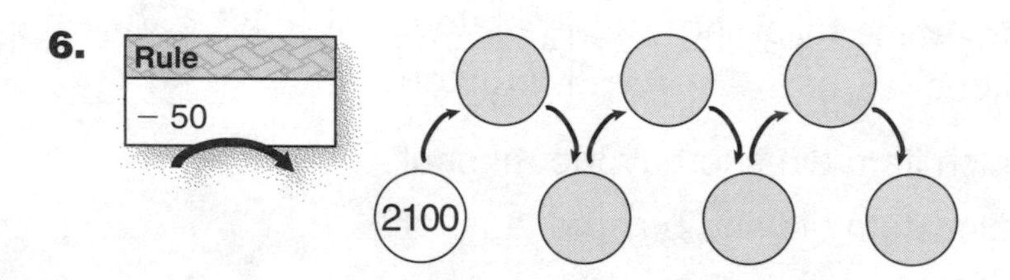

7.

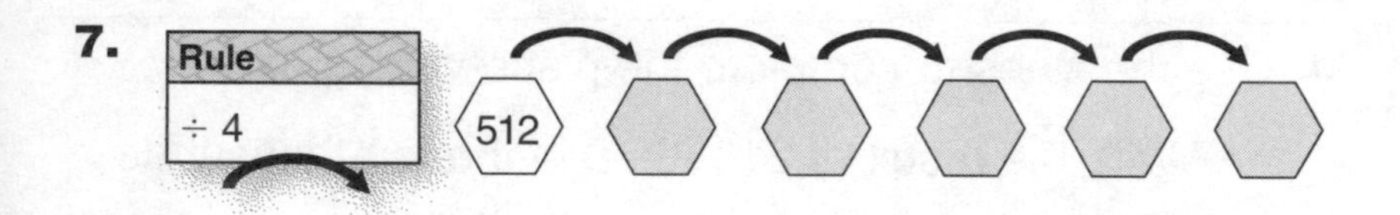

8.

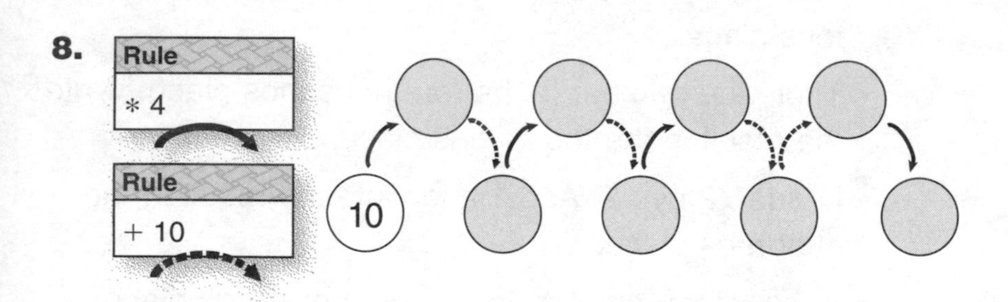

9.

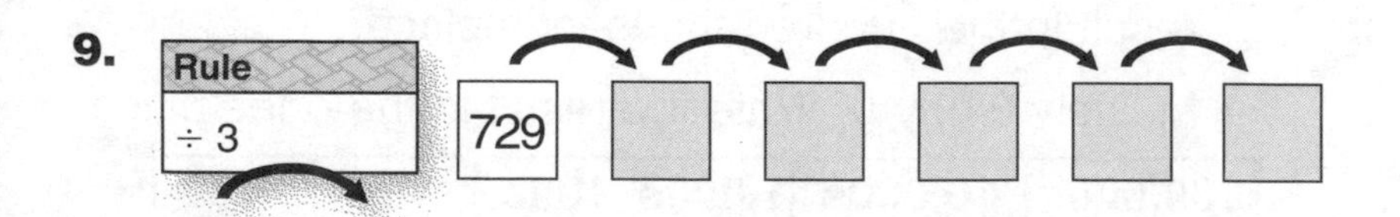

Use with or after Lesson 4.3.

Practice Set 25

Write all of your answers on a separate sheet of paper.

Add or subtract.

1. 15.3 + 12.9

2. 5 − 1.6

3. 11.3 − 6

4. 8.24 + 3.9

5. 2.09 + 2.2

6. 1.9 − 0.41

7. 5.15 − 3.67

8. 0.69 + 0.75

9. 18.5 + 3.9

10. 1.7 − 1.34

Write the words for the following numbers.

11. 296,069

12. 1,312,743

13. 854.09

14. 3,969,231

15. 8,201,774

Complete the "What's My Rule?" tables.

16.

Rule
out = in * 200

in	out
7	1,400
9	
12	
14	
35	

17.

Rule

in	out
7	$14\frac{1}{2}$
10	$17\frac{1}{2}$
	20
$13\frac{1}{2}$	
$22\frac{1}{2}$	30

Write all of your answers on a separate sheet of paper.

18. Four people are going to share $68 equally.

a. How many $10 bills does each person get?

b. How many dollars are left to share?

c. From the money that remains, how many $1 bills does each person get?

d. What is the total number of dollars each person gets?

e. Number model: $4 * 17 = ■$

19. Seven people are going to share $112 equally.

a. How many $100 bills does each person get?

b. From the money that remains, how many $10 bills does each person get?

c. How many dollars are left to share?

d. From the money that remains, how many $1 bills does each person get?

e. What is the total number of dollars each person gets?

f. Write a number model for this problem.

20. Six people are going to share $681 equally.

a. How many $100 bills does each person get?

b. How many dollars are left to share?

c. From the money that remains, how many $10 bills does each person get?

d. How many dollars are left to share?

e. From the money that remains, how many $1 bills does each person get?

f. How many dollars are left over?

Practice Set 26

Write all of your answers on a separate sheet of paper.

Solve.

1. \$17.03 − \$3.85

2. \$8.22 + \$30

3. \$6.05 + \$2.80

4. \$16.23 − \$9.66

5. \$14.85 − \$12.60

6. \$5.99 + \$2.49

7. \$20 + \$16.56

8. \$15.68 − \$5.97

9. \$7.48 + \$21.70

10. \$50 − \$19.79

Complete the "What's My Rule?" tables.

11.

Rule
out = in * 12

in	out
3	36
5	
7	
11	
15	

12.

Rule

in	out
8	24
14	30
	44
35	
43	59

Make a name-collection box for each number listed below. Use as many different numbers and operations as you can.

Example

419
$(\frac{1}{5} * 2{,}500) - (9 * 9)$
(205 * 2) + 9
838 ÷ 2
(500 − 85) + 4

13. 380

14. 176

15. 4,218

16. 510

17. 6,111

18. 495

Practice Set 27

Write all of your answers on a separate sheet of paper.

Measure each line segment to the nearest centimeter. Record the measurements in centimeters and meters.

1. ______________ **2.** ____

3. ______________________

4. ____________________________________

5. ___________________________

For each number sentence write *T* if it is true or *F* if it is false.

6. $92 - 40 = 42$ **7.** $80 = 18 + 62$

8. $9 * 7 = 67$ **9.** $6 > 72 / 9$

10. $500 + 80 < 550$ **11.** $448 - 15 > 400$

Use digits to write the following numbers.

12. two hundred sixty thousand, four hundred fifty-three

13. two hundred eighty-six and thirty-eight hundredths

14. three hundred fourteen thousand, six hundred ninety-one

15. one million, seventy-four thousand, nine hundred sixty-eight

16. six million, seven hundred nine thousand, eight hundred forty-five

Practice Set 27 (cont.)

Write all of your answers on a separate sheet of paper.

Solve.

17. 180 ÷ 6 = ■

18. 24 * 6 = ■

19. 810 = 90 * ■

20. 40 * 70 = ■

21. 80 * ■ = 3,200

22. ■ ÷ 100 = 6

23. 8,400 ÷ 700 = ■

24. 36 * ■ = 72

25. ■ ÷ 5 = 70

26. 9 * 200 = ■

27. 5,400 ÷ ■ = 9

28. 12 * 120 = ■

29. How much money, without tax, will I need for 4 VCR tapes that cost $3.25 each?

30. How many dollars are in 22 five-dollar bills?

31. If 1 block is 200 meters long, how far will you run in 23 blocks?

32. There are about 8 blocks in one mile. How many blocks are in 5 miles?

33. There are 5,280 feet in one mile. How many yards are in one mile? (Reminder: 1 yd = 3 ft)

34. How many five-dollar bills are in $325.00?

Practice Set 28

Write all of your answers on a separate sheet of paper.

Complete.

1. 500 cm = ___ m

2. ___ cm = 8.1 m

3. ___ mm = 72 cm

4. 150 cm = ___ m

5. 0.35 m = ___ cm

6. 63 cm = ___ m

7. ___ mm = 9.8 cm

8. ___ m = 375 cm

Solve.

9. $\begin{array}{r} 1,800 \\ -\ 927 \\ \hline \end{array}$

10. $\begin{array}{r} 3,684 \\ -\ 485 \\ \hline \end{array}$

11. $\begin{array}{r} 3,164 \\ +\ 5,791 \\ \hline \end{array}$

12. $\begin{array}{r} 8,261 \\ -\ 3,540 \\ \hline \end{array}$

13. $\begin{array}{r} 600 \\ -\ 31 \\ \hline \end{array}$

14. $\begin{array}{r} 475 \\ +\ 250 \\ \hline \end{array}$

15. $\begin{array}{r} 1,834 \\ +\ 8,365 \\ \hline \end{array}$

16. $\begin{array}{r} 469 \\ -\ 70 \\ \hline \end{array}$

17. $\begin{array}{r} 1,200 \\ -\ 30 \\ \hline \end{array}$

18. $\begin{array}{r} 2,444 \\ -\ 382 \\ \hline \end{array}$

19. $\begin{array}{r} 729 \\ +\ 682 \\ \hline \end{array}$

20. $\begin{array}{r} 1,356 \\ -\ 1,172 \\ \hline \end{array}$

21. $\begin{array}{r} 4,321 \\ -\ 1,234 \\ \hline \end{array}$

22. $\begin{array}{r} 500 \\ -\ 42 \\ \hline \end{array}$

23. $\begin{array}{r} 1,300 \\ -\ 485 \\ \hline \end{array}$

Practice Set 29

Write all of your answers on a separate sheet of paper.

1. Use the clues to complete the place-value puzzle.

- Add 43 and 23. Divide by 11 and write the result in the thousandths place.
- Triple the number in the thousandths place and divide by 2. Write the result in the hundredths place.
- Multiply 8 ∗ 9. Subtract 68. Write the result in the ones place.
- Subtract the number in the hundredths place from 57 and divide by 6. Write the result in the hundreds place.
- Divide 36 by the number in the ones place. Write the result in the tens place.
- Subtract the number in the tens place from the number in the hundredths place. Write the result in the tenths place.

100s	10s	1s		0.1s	0.01s	0.001s
			.			

Write the missing numbers.

2.

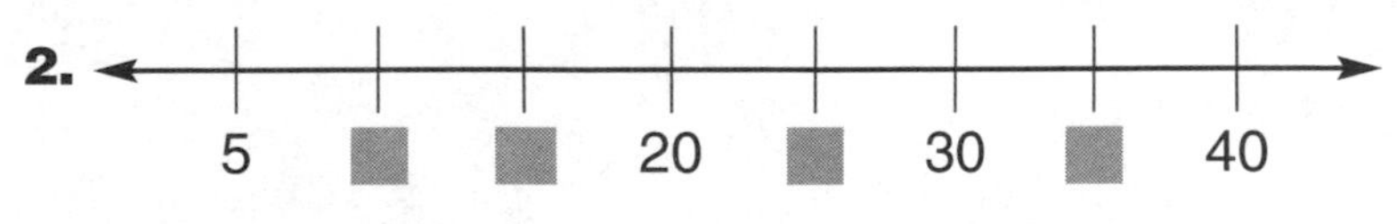

3.

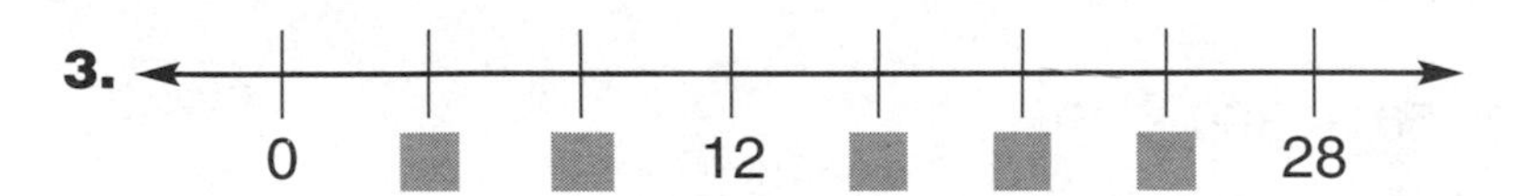

Practice Set 29 (cont.)

Write all of your answers on a separate sheet of paper.

Complete the "What's My Rule?" tables.

4.

Rule

in	out
10	18.25
12.5	
	22.75
54.10	
88.0	96.25

5.

Rule
out = in ÷ 3

in	out
600	200
900	
1,200	
1,500	
1,800	

Rewrite the number sentences with parentheses to make them correct.

6. 26 ÷ 2 − 7 = 6

7. 41.2 = 7 ∗ 6 − 0.8

8. 130 − 15 − 60 = 55

9. 118 = 2 ∗ 55.7 + 3.3

10. 10 ∗ 2.1 + 5 ∗ 12.2 = 82

11. 30 = 6 ∗ 20 − 90

12. 11 ∗ 12 + 7 − 4 = 165

13. 99 = 11 ∗ 50 − 41

14. 50 + 300 ÷ 5 = 70

15. 200 ∗ 2.6 − 1.1 + 5 = 305

Practice Set 30

Write all of your answers on a separate sheet of paper.

Solve.

1. 4 * 80

2. 60 * 50

3. 20 * 7

4. 5 * 40

5. 70 * 60

6. 30 * 9

7. 80 * 20

8. 40 * 80

9. 60 * 6

10. 90 * 80

Find the missing factors.

11. 30 * ___ = 2,700

12. 90 * ___ = 450

13. 5 * ___ = 250

14. ___ * 2 = 120

15. 70 * ___ = 2,800

16. 30 * ___ = 600

17. ___ * 50 = 4,000

18. 80 * ___ = 640

Write the missing numbers.

19.

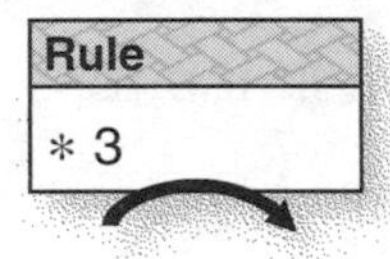

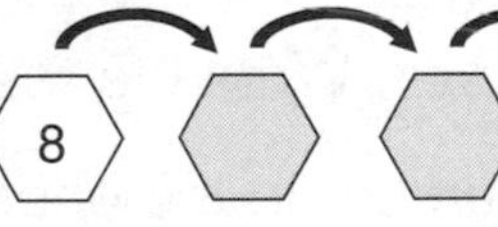

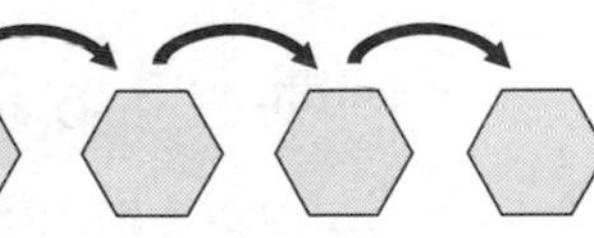

20.

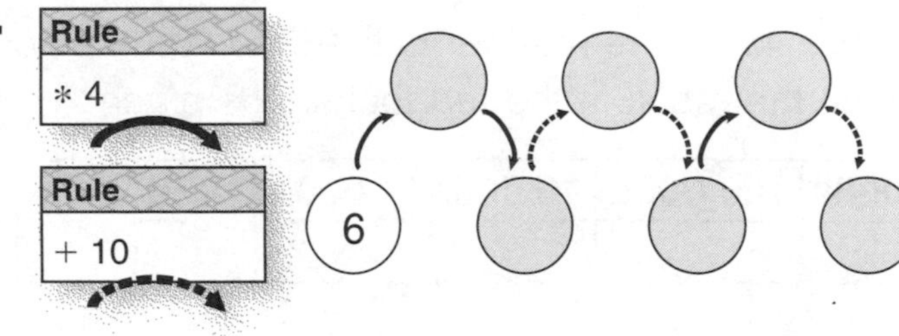

Write all of your answers on a separate sheet of paper.

21.

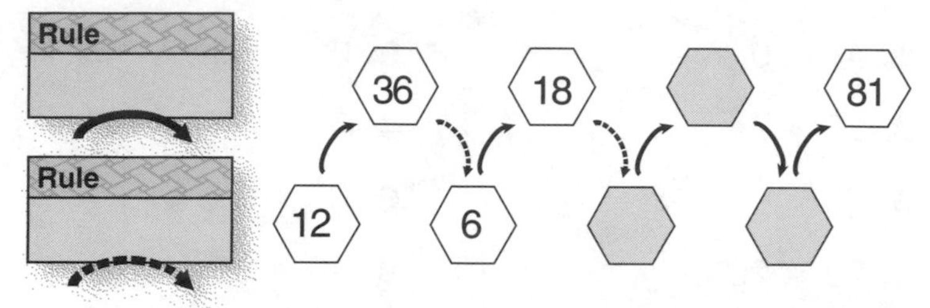

22. Use the clues to complete the place-value puzzle.

- Divide 60 by 12. Write the result in the ones place.
- Triple the number in the ones place and divide by 3. Write the result in the hundreds place.
- 4 is the square of ■. Write the result in the millions place.
- Halve the number in the millions place. Multiply by 6 and write the result in the thousands place.
- Multiply 3 by itself. Write the result in the tens place.
- Subtract the number in the ones place from the number in the thousands place. Write the result in the hundred-thousands place.
- Divide 42 by the number in the thousands place. Write the result in the ten-thousands place.

1,000,000s	100,000s	10,000s	1,000s	100s	10s	1s

Practice Set 31

Write all of your answers on a separate sheet of paper.

Estimate the sums using thousands and hundreds.

1.	**2.**	**3.**	**4.**
2,860	1,010	5,272	7,280
1,452	1,345	2,470	6,109
+ 9,024	+ 6,813	+ 1,391	+ 6,109

5.	**6.**	**7.**	**8.**
3,554	9,656	2,843	6,590
7,490	1,970	7,154	4,483
6,708	6,213	1,041	3,217
+ 2,143	+ 3,160	2,630	1,670
		+ 1,274	+ 2,532

Solve.

9. You went to the store with a \$10 bill and a \$5 bill. Your groceries cost \$12.36. How much change should you get?

10. $\frac{20}{4}$ **11.** $\frac{36}{9}$ **12.** $7 * 7$ **13.** $6 * 6$

14. $9 * 80$ **15.** $12 + 13$ **16.** $41 - 25$ **17.** $7 * 8$

18. $5 * 6$ **19.** $\frac{12}{12}$ **20.** $\frac{40}{10}$ **21.** $\frac{48}{6}$

Use digits to write the following numbers:

22. forty-five thousand, three hundred ninety-two

23. four hundred fifty-nine thousand, seven hundred three

Practice Set 32

Write all of your answers on a separate sheet of paper.

Estimate the answer. Write a number model to show how you estimated.

1. There are 12 cans in each case. How many cans in 37 cases?
2. A large can of peaches holds 112 ounces. How many ounces are in 6 cans?
3. Ellen uses 1 gallon of lemonade to serve 22 people. How many people could she serve with 53 gallons of lemonade?

Use the statements below to help you solve the problems.

- The average person throws away about 5 pounds of trash per day.
- One ton is equal to 2,000 pounds.
- There are about 250 million people in the United States.

4. How much trash does the average person throw away in one week?
5. How much trash does the average person throw away in one year?
6. About how many tons is that?
7. About how many tons of trash does the average family of 4 throw away in one year?
8. Does the population of the United States produce more or less than 10 million tons of trash per year?
9. About how many tons of trash does the United States produce in one year?

Use with or after Lesson 5.4.

Practice Set 32 (cont.)

Write all of your answers on a separate sheet of paper.

Make a name-collection box for each number listed below. Use as many different numbers and operations as you can.

Example

24
XXIV
$48 \div 2$
$29 - 5$
12×2

10. 15

11. 100

12. 54

13. 73

True or False?

Example 12 + 15 = 25 *(False)*
4 * (3 + 1) = 16 *(True)*

14. 7 * 9 = 54

15. 3 * (4 + 5) = 27

16. 5 * 6 = 40

17. (99 + 13) = 103

---

18. Use the clues to complete the place-value puzzle.

- Divide 18 by 6. Write the result in the ones place.
- Double the number in the ones place. Divide by 3. Write the result in the tens place.
- Write the result of 8 * 5 divided by 10 in the thousands place.
- Multiply 7 by 2. Subtract 7. Write the result in the hundreds place.

1,000s	100s	10s	1s

Practice Set 33

Write all of your answers on a separate sheet of paper.

Use partial products to multiply.

1. $47 * 6$

2. $89 * 5$

3. $27 * 13$

4. $36 * 46$

5. $90 * 19$

6. $23 * 62$

7. $159 * 41$

8. $613 * 18$

Solve.

9. $23 * 4$

10. $18 \div 1$

11. $54 + 36$

12. $9 * 35$

13. $180 / 10$

14. $78 - 23$

15. $50 * 60$

16. $162 / 6$

17. $48 - 12$

Examine the data sets. Find the mean of each.

Example 7 6 5 9 8

Step 1 Find the total of the numbers in the data set.

$7 + 6 + 5 + 9 + 8 = 35$

Step 2 Count how many numbers are in the data set.

There are 5 numbers in all.

Step 3 Divide the total by the amount of numbers.

$\frac{35}{5} = 7$ Mean = 7

18. 5 8 8 11 7 9

19. 12 10 8 14 11

20. 3 2 3 1 6 4 2

21. 14 15 18 13

Practice Set 33 (cont.)

Write all of your answers on a separate sheet of paper.

Who am I?

22. Clue 1: I am less than 10.
Clue 2: I am an even number.
Clue 3: I am a square number.

23. Clue 1: I am less than 100.
Clue 2: The sum of my digits is 17.
Clue 3: I am an even number.

24. Clue 1: I am a number between 100 and 200.
Clue 2: The sum of my digits is 3.
Clue 3: My ones digit is two times my hundreds digit.

25. The first figure is $\frac{3}{4}$ of the whole. What fraction of the whole is each of the other figures?

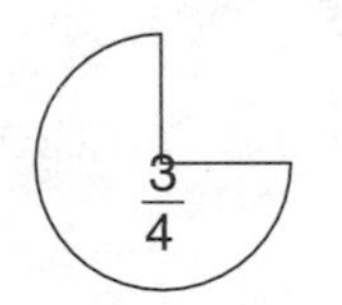

a. 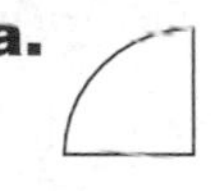

b. 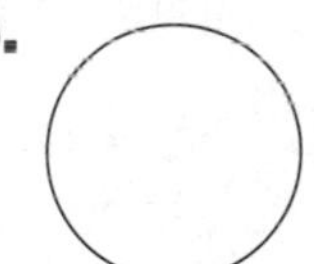

c. 

26. Write a number for each picture. Use 0 or $\frac{0}{4}$, $\frac{1}{4}$, $\frac{1}{2}$, $\frac{3}{4}$, and $\frac{4}{4}$ or 1.

a. 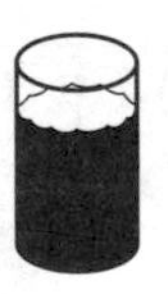

b. 

c. 

d. 

e. 

Practice Set 34

Write all of your answers on a separate sheet of paper.

Use the lattice method to find the following products.

1. 52 ∗ 9

2. 14 ∗ 6

3. 58 ∗ 16

4. 69 ∗ 24

5. 23 ∗ 87

6. 42 ∗ 291

7. 61 ∗ 107

8. 35 ∗ 463

---

9. Use the clues to complete the place-value puzzle.

- Write the result of 36 – 28 in the thousands place.
- Multiply 6 ∗ 3 and subtract 17. Write the result in the ones place.
- Triple the number in the thousands place and divide by 4. Write the result in the hundreds place.
- Subtract 3 from the result of 270 divided by 90. Write the result in the ten-thousands place.
- Divide 135 by 27. Write the result in the hundred-thousands place.
- Subtract the number in the hundred-thousands place from the number in the thousands place. Write the result in the tens place.

100,000s	10,000s	1,000s	100s	10s	1s

Practice Set 34 (cont.)

Write all of your answers on a separate sheet of paper.

Solve.

10. 326 * 30

11. 965 − 86

12. 541 − 8

13. 160 + 1,400

14. 6,045 + 248

15. 2,289 + 1,374

16. 18 * 11

17. 4,371 + 8,148

18. 890 − 15

19. (70 + 15) * 4

20. 67 − (8 * 4)

21. 430 + 70 + 145

22. (72 / 8) * 3

23. Alice records her weight change every week. For the past three weeks she recorded + 3, −1, and + 2 pounds. Can you tell how much she weighs?

Rewrite the number sentences with parentheses to make them correct.

24. 412 = 70 * 5 + 1 − 8

25. 6 * 10 − 5 = 55

26. 81 = 7 + 2 * 9

27. 39 − 16 − 4 = 19

28. 44 − 13 + 23 = 8

29. 8 * 5 + 2 * 18 = 76

30. 135 = 9 * 9 + 6

31. 4 * 3 + 7 * 6 = 240

Practice Set 35

Write all of your answers on a separate sheet of paper.

Use digits to write the following numbers.

1. two million, seven hundred eighteen thousand, nine hundred twenty

2. seven hundred sixty-nine thousand, two hundred thirty-one

3. eighteen and nine hundred seventy-eight thousandths

Write the words for the following numbers.

4. 18,564,290

5. 48.128

6. 5,773,963

7. 102,756

Write the amounts.

8. $1 Q Q Q Q Q Q Q D P P P P

9. $5 $1 $1 Q Q Q N N N N

10. $100 $20 $5 $5 $5 $1 $1

11. Q Q Q Q Q Q D N N N P P

Practice Set 36

Write all of your answers on a separate sheet of paper.

Write the numbers in order from least to greatest.

1. 10,000; 10^5; 10 * 10; 1 thousand

2. 1 million; 10^4; 10 [hundreds]; 10 * 10 * 10 * 10 * 10

3. 100,000; 10^2; 10 [thousands]; 10 * 10 * 10

4. 10 [tenths]; 1 thousand; 10^5; 10 * 10

Solve.

5. 20 ÷ 4 **6.** 36 / 9 **7.** 12 ÷ 12 **8.** 40 ÷ 10

9. 48 ÷ 6 **10.** 8 / 4 **11.** 0 / 4 **12.** 18 ÷ 1

13. 12 ÷ 4 **14.** 24 / 8 **15.** 4 ÷ 8 **16.** 144 ÷ 12

Solve.

17. 35 * 4

18. 62 * 3

19. 265 * 8

20. 43 + 18

21. 710 + 136

22. 500 + 4,500

23. 3,249 − 1,933

24. 1,865 − 674

Write all of your answers on a separate sheet of paper.

Solve the following problems. (The prices include tax.)

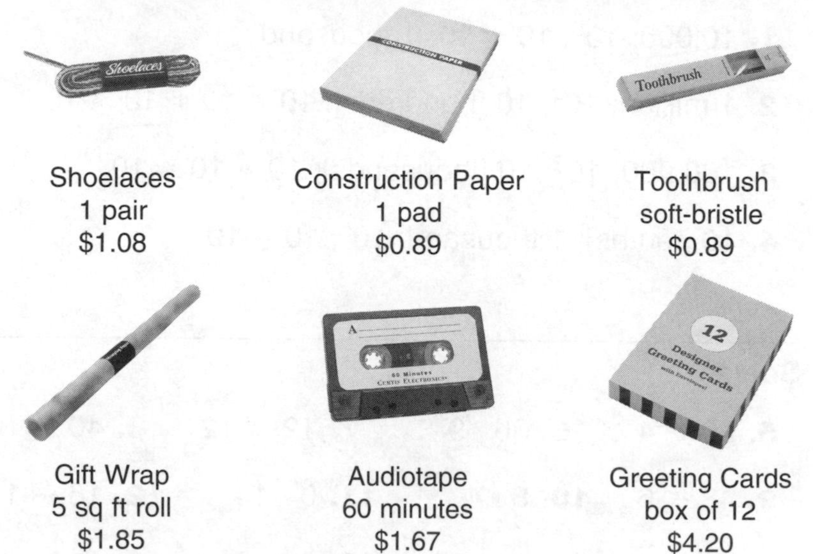

Shoelaces
1 pair
$1.08

Construction Paper
1 pad
$0.89

Toothbrush
soft-bristle
$0.89

Gift Wrap
5 sq ft roll
$1.85

Audiotape
60 minutes
$1.67

Greeting Cards
box of 12
$4.20

25. Tom is going shopping. Each of his two children needs a pair of shoelaces, a toothbrush, and a pad of construction paper. Estimate to the nearest dollar how much money he will need.

26. Brianna needs enough wrapping paper to cover her five presents. She has 6 dollars and she estimates that she will need three rolls. Does she have enough money? What is the difference between the amount of money she has and the amount she needs?

27. Joe and Ahmed are researching the 50-year history of their school. They need to buy enough audiotapes for an hour-and-a-half interview with the principal. How much money do they need?

28. How much money does each of the greeting cards in the box of 12 cost?

Practice Set 37

Write all of your answers on a separate sheet of paper.

Find the missing number for each Fact Triangle. Write the fact family for that triangle.

Example

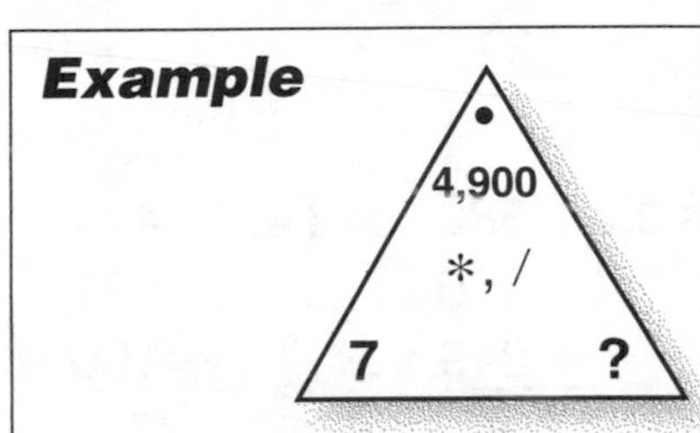

Missing Number: 700

Fact family: $7 * 700 = 4{,}900$
$700 * 7 = 4{,}900$
$4{,}900 / 7 = 700$
$4{,}900 / 700 = 7$

1.

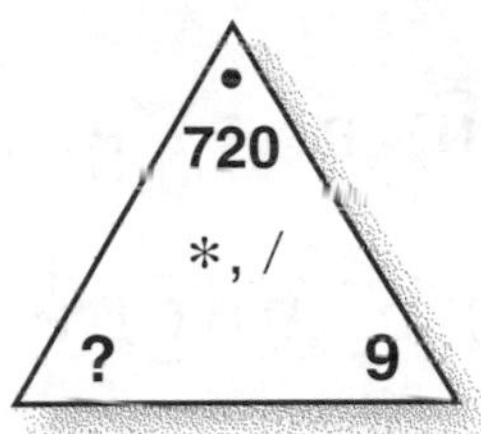

2.

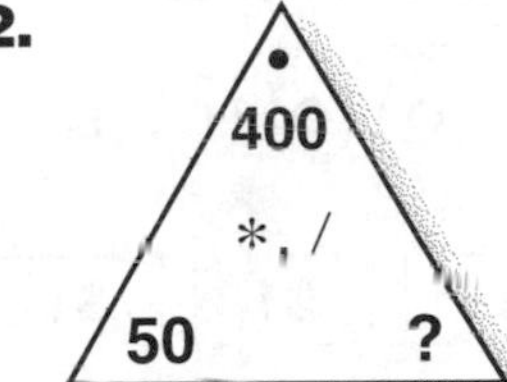

3.

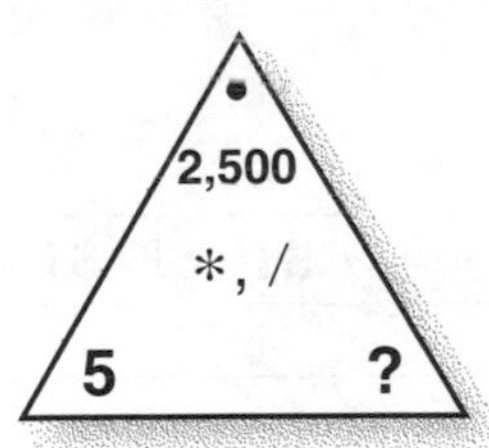

4.

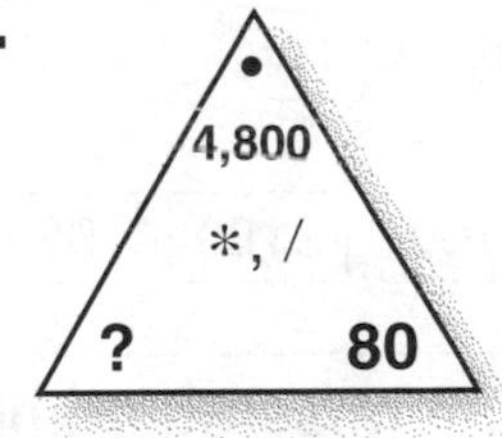

5.

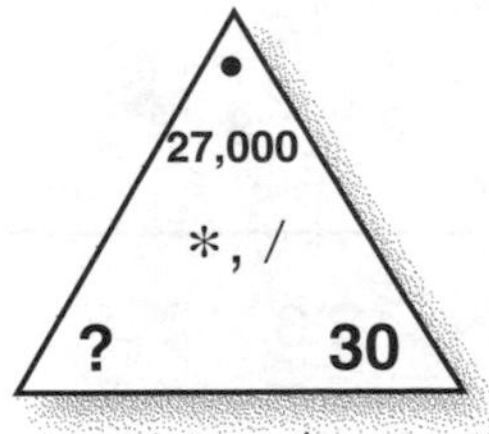

6.

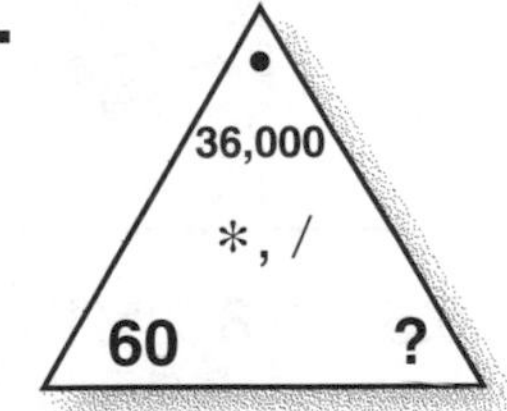

Practice Set 37 (cont.)

Write all of your answers on a separate sheet of paper.

Solve.

7. $3,641 - 2,040$

8. $21 * 9$

9. $62 * 21$

10. $178 * 5$

11. $408 + 323 + 475$

12. $205 + 335 + 182$

13. $382 + 416 + 249$

14. $414 + 627 + 100$

Write the amounts.

15. Q Q Q Q Q D D N N P P P

16. \$1 \$1 \$1 Q D D D D P P

17. \$5 \$5 \$5 \$5 \$5 \$1 Q N N

18. \$100 \$100 \$20 \$20 \$5 \$1 \$1

Write the missing numbers.

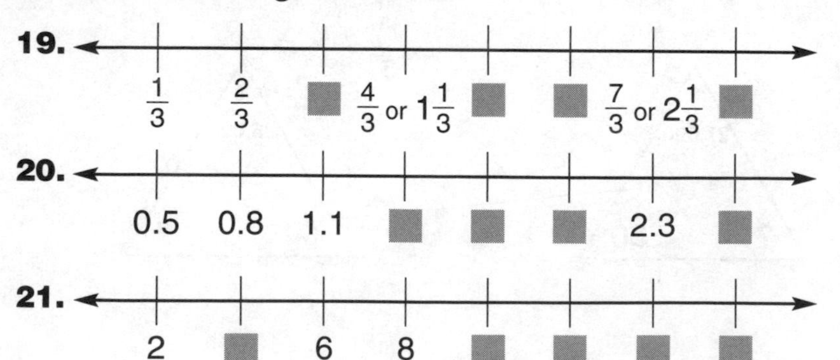

Practice Set 38

Write all of your answers on a separate sheet of paper.

Use the partial-quotients method to divide.

1. $4\overline{)124}$

2. 154 ÷ 6

3. 355 / 9

4. 247 ÷ 11

5. $3\overline{)195}$

6. $5\overline{)256}$

7. 129 / 14

8. 197 / 8

9. 232 ÷ 12

10. 158 ÷ 5

11. $7\overline{)164}$

12. 235 / 16

---

13. Eight people are going to share $168 equally.

a. How many $10 bills does each person get?

b. How many dollars are left to share?

c. From the money that remains, how many $1 bills does each person get?

d. What is the total number of dollars each person gets?

e. Write a number model for this problem.

14. Five people are going to share $1,025 equally.

a. How many $100 bills does each person get?

b. From the money that remains, how many $10 bills does each person get?

c. How many dollars are left to share?

d. From the money that remains, how many $1 bills does each person get?

e. What is the total number of dollars each person gets?

f. Write a number model for this problem.

Practice Set 38 (cont.)

Write all of your answers on a separate sheet of paper.

Solve.

15. $352 - 247$

16. $118 * 7$

17. $3{,}276 + 1{,}398$

18. $768 - 89$

In each set of problems below, do as many exercises as you can in one minute. Ask someone to time you.

Problem Set 1

19. 12 − 6 = ■

20. 16 ÷ 4 = ■

21. 8 * ■ = 40

22. 54 / 9 = ■

23. 5 + 3 = ■

24. 11 − 8 = ■

25. 100 ÷ 10 = ■

26. ■ * 9 = 36

27. 12 * 6 = ■

28. 3 * ■ = 27

29. 4 + 7 = ■

30. 20 / 5 = ■

31. 15 − 8 = ■

32. 6 + 9 = ■

33. 36 ÷ ■ = 6

Problem Set 2

34. 9 + 2 = ■

35. 32 / 8 = ■

36. 5 * ■ = 25

37. 30 ÷ 5 = ■

38. 6 + ■ = 14

39. 10 − 7 = ■

40. 64 ÷ 8 = ■

41. ■ * 7 = 56

42. 4 * 6 = ■

43. 4 * ■ = 48

44. 16 − 7 = ■

45. 45 ÷ 5 = ■

46. 16 / 4 = ■

47. 12 − ■ = 6

48. 7 + 6 = ■

Practice Set 39

Write all of your answers on a separate sheet of paper.

Estimate. Tell whether the answer will be in the tens, hundreds, or thousands.

1. 951 ÷ 8
2. 165 / 9
3. 677 * 2
4. 92 ÷ 6
5. 924 ÷ 5
6. 472 / 15
7. 67 * 12
8. 762 ÷ 31
9. 389 ÷ 7
10. 437 ÷ 3
11. 32 * 65
12. 135 * 5

Solve.

13. Four friends shared 516 trading cards. How many trading cards did each person get?

14. Alec worked 6 days and earned $354. How much did he earn per day?

15. Meredith has a collection of glass animals. She stores her collection in 3 boxes. Each box holds 24 glass animals. How many glass animals are in her collection?

16. To raise money, the nature club sold 92 boxes of note cards. Each box sold for $6. How much money did the club raise?

Complete the "What's My Rule?" tables.

17.

Rule
out = in * 22

in	out
3	66
4	
8	
14	
16	

18.

Rule

in	out
12	132
13	143
	154
18	
20	220

Practice Set 39 (cont.)

Write all of your answers on a separate sheet of paper.

19. How many pieces of fruit are there?

20. What fraction of the fruit is apples?

21. What fraction of the fruit is pears?

22. What fraction of the fruit is bananas?

23. What fraction of the fruit is oranges?

Complete the frames-and-arrows problem.

24.

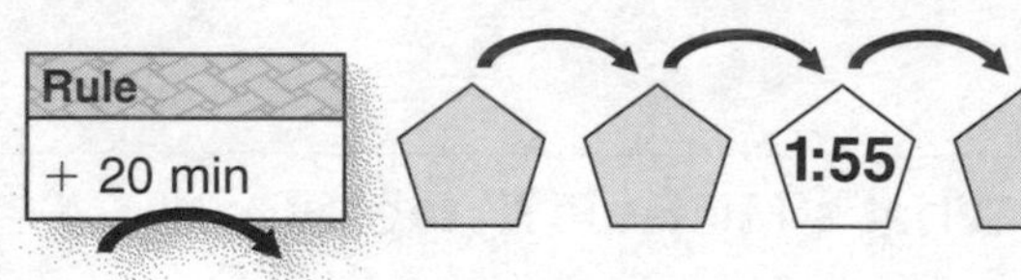

Find the missing factors.

25. 8 * ■ = 24

26. ■ * 90 = 360

27. ■ * 7 = 49

28. 2 * ■ = 960

29. 60 * ■ = 3,600

30. ■ * 7 = 350

31. 8 * ■ = 640

32. ■ * 32 = 640

Practice Set 40

Write all of your answers on a separate sheet of paper.

Write each answer as a mixed number and as a decimal.

1. 90 ÷ 8

2. 169 ÷ 5

3. 183 ÷ 10

4. 297 ÷ 4

5. 65 ÷ 2

6. 93 ÷ 6

Divide.

7. A set of 4 chairs costs $125. What is the cost per chair?

8. Emma is making a quilt. She cuts 6 squares from each square foot of fabric. She needs 100 squares. How many square feet of fabric does she need?

Solve.

9. 20 * 80

10. 16 * 10

11. 82 * 100

12. 7 * 300

13. 91 * 10

14. 7.6 * 100

15. 14 * 200

16. 75 * 60

17. 400 * 5.0

18. 30.4 * 10

19. 1.9 * 200

20. 19 * 200

21. How many 7s in 1,400?

22. How many 70s in 4,900?

Practice Set 41

Write all of your answers on a separate sheet of paper.

Write the number of degrees the minute hand moves.

1. from 1:00 to 1:15

2. from 2:00 to 2:30

3. from 11:00 to 11:03

4. from 7:00 to 7:25

5. from 10:00 to 10:25

6. from 1:00 to 2:00

Complete the frames-and-arrows problems.

7.

Rule
$*$ 10

50 500

8.

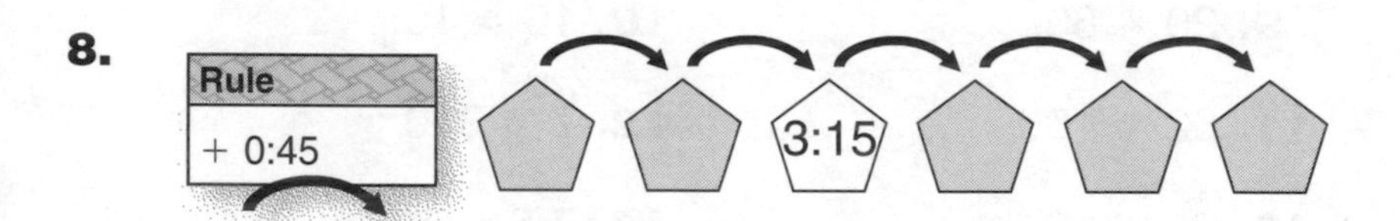

9.

Rule
+ 10¢

Rule
− $1.80

$3.50 10¢

Use with or after Lesson 6.8.

Practice Set 41 (cont.)

Write all of your answers on a separate sheet of paper.

Solve. (Prices include tax.)

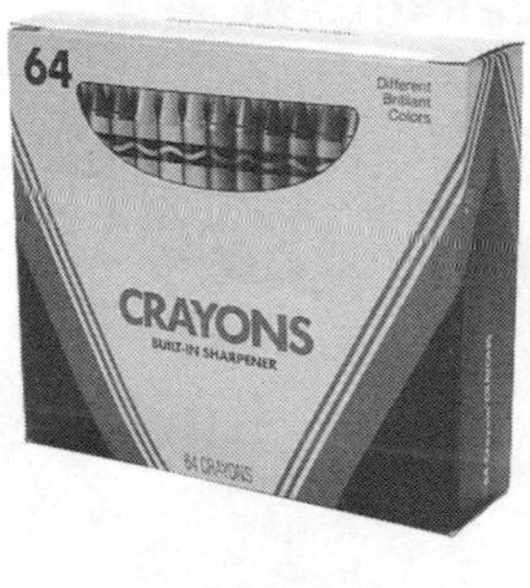

$3.99 $1.99 $1.59

10. Ms. Jackson wants to buy enough crayons to give 1 to each of her 29 students. She has $3.50.

a. What can she buy?

b. How many crayons will she have left over?

11. How many boxes of 16 crayons would it take to equal the number in the 64-crayon box?

12. How much would this cost?

13. Estimate whether $18 is enough to buy 5 boxes of 64 crayons.

Solve.

14. 29 * 3

15. 57 * 8

16. 495 * 6

17. 307 * 4

18. 860 * 7

19. 334 * 11

20. Draw an array that represents the number model 4 * 7 = 28

Practice Set 42

Write all of your answers on a separate sheet of paper.

Judy brought 12 quarters to the arcade. She spent $\frac{1}{3}$ of them on video games and $\frac{1}{2}$ on basketball.

1. How much did she spend on video games?

2. How much did she spend on basketball?

3. How much money was left?

4. What fraction of the total is that?

Who am I?

5. Clue 1: I am less than 20.
Clue 2: I am an odd number.
Clue 3: I am the square root of 225.

6. Clue 1: I am less than 100.
Clue 2: I am an even number.
Clue 3: I can be represented by an array of 7 rows and 8 columns.

7. Clue 1: I am less than 1,000.
Clue 2: The sum of my digits is 26.
Clue 3: My ones digit is eight.

Use with or after Lesson 7.2.

Practice Set 42 (cont.)

Write all of your answers on a separate sheet of paper.

Complete the "What's My Rule?" tables.

8.

Rule
out = in * 25

in	out
3	
4	
8	
	350
16	

9.

Rule
out = in / 9

in	out
81	
54	
	12
117	
	7

Solve.

10. 207 − ■ = 65

11. 521 + 227 = ■

12. 190 + 448 = ■

13. 690 − 237 = ■

14. 1,416 − 948 = ■

15. 60 = 5,400 ÷ ■

16. 869 − 44

17. 483 − 355

18. 68 * 8

19. 8,521 + 4,349

20. 1,050 * 4

21. 2,756 + 1,711

Practice Set 43

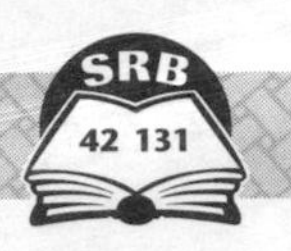

Write all of your answers on a separate sheet of paper.

Write the fraction of the figure that is shaded.

1. 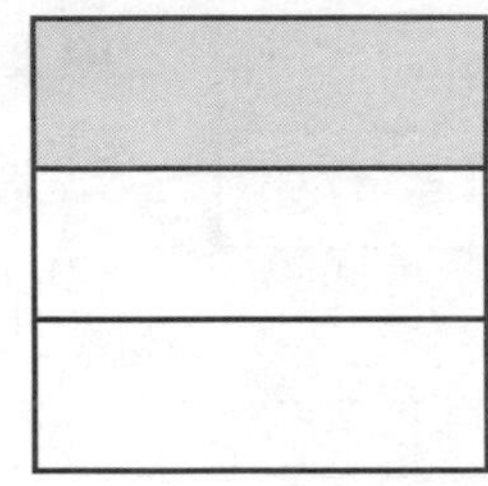

2. 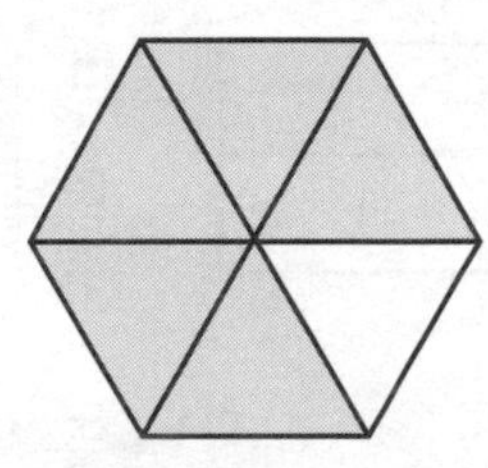

3. 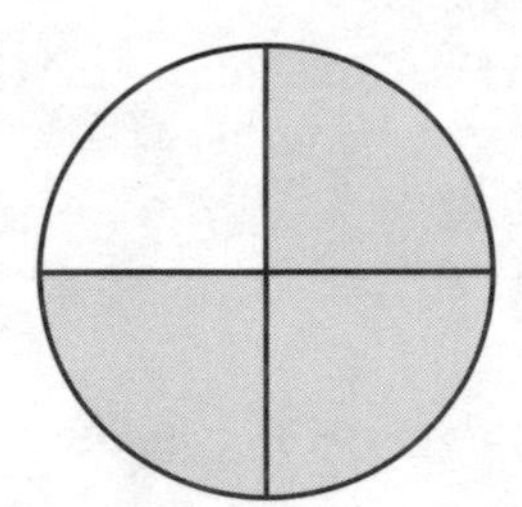

4. 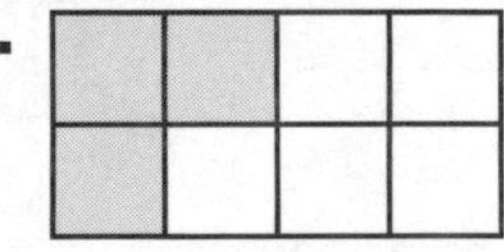

5. 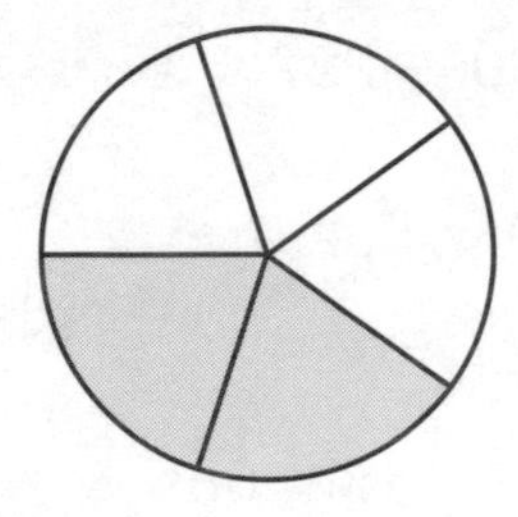

6. 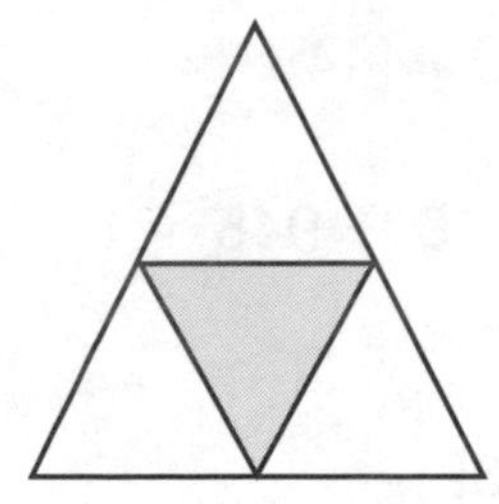

Write < or > to make each number sentence true.

7. 9,608 ■ 9,906

8. 48,549 ■ 48,459

9. 113,012 ■ 131,102

10. 278,300 ■ 79,309

11. 3,780,576 ■ 420,777

12. 50,701,318 ■ 5,710,381

Practice Set 43 (cont.)

Write all of your answers on a separate sheet of paper.

Divide. Write the answer as a mixed number.

13. 41 ÷ 6

14. 61 / 3

15. 87 / 4

16. 149 ÷ 2

17. 268 / 5

18. 731 ÷ 8

19. 680 ÷ 9

20. 425 / 7

The residents of an apartment building were asked how many people live in their households. The tallies in the table show the results of the survey. Use the table to help you answer the questions below.

21. How many households were interviewed?

22. How many people live in the building?

23. What is the median number of people per household in the building?

24. What is the approximate mean number (average) of people per household in the building to the nearest whole number?

People per household	Number of households
1	///
2	~~////~~
3	~~////~~ //
4	~~////~~ /
5	//
6	//

Practice Set 44

Write all of your answers on a separate sheet of paper.

Solve.

1. A cake was cut into eight pieces.

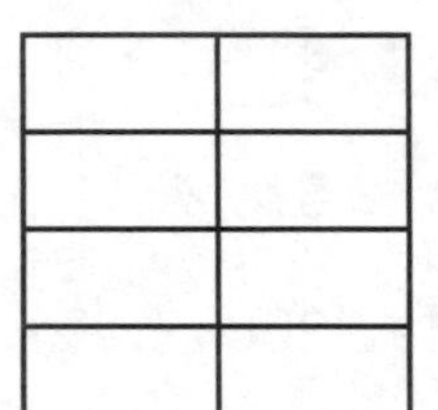

a. Monday night $\frac{5}{8}$ of the cake was eaten. How much was left?

b. Marilyn ate $\frac{1}{4}$ of the cake and Billy ate $\frac{1}{8}$ of the cake. How much cake did they eat in all?

2. Rita draws a line segment $2\frac{1}{2}$ inches long. Then she erases $\frac{3}{8}$ inch. How long is the line segment now?

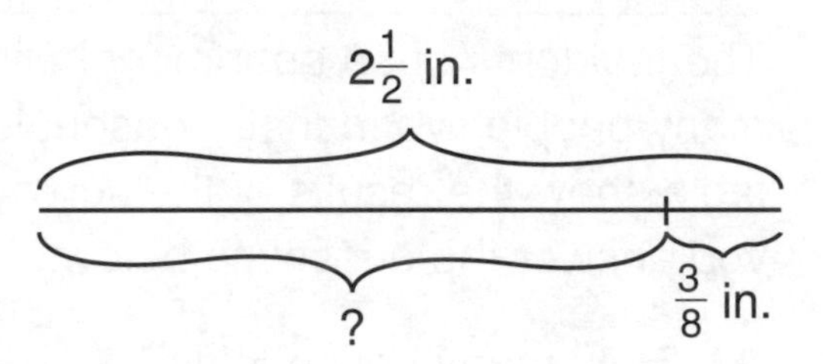

Write two multiplication and two division problems for each of the following Fact Triangles:

3.

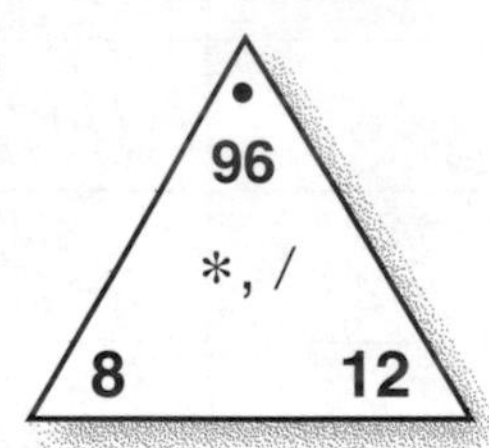

4.

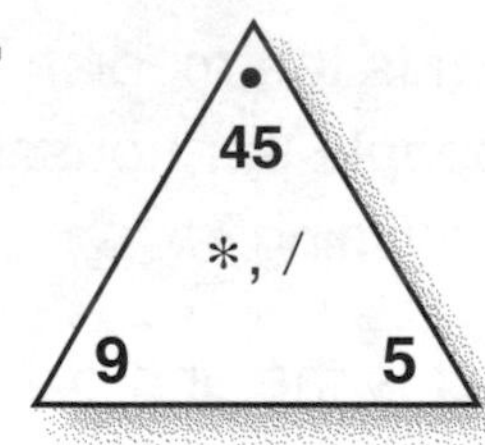

5.

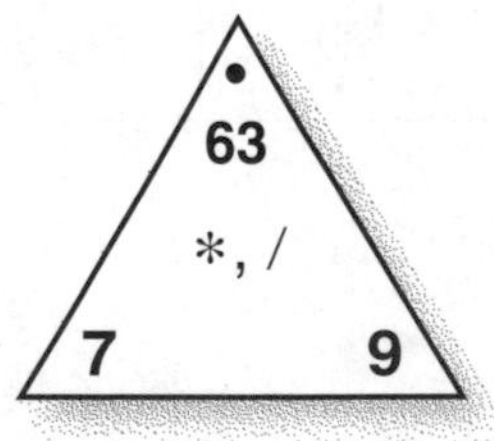

6.

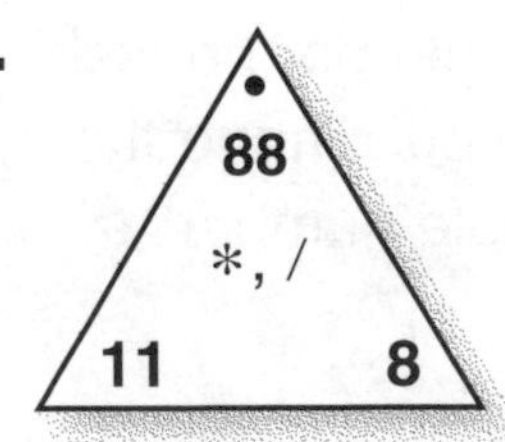

Practice Set 44 (cont.)

Write all of your answers on a separate sheet of paper.

Solve.

7. $\begin{array}{r} 16 \\ * \ 24 \\ \hline \end{array}$

8. $\begin{array}{r} 91 \\ - \ 35 \\ \hline \end{array}$

9. $\begin{array}{r} 31 \\ * \ 42 \\ \hline \end{array}$

10. $7\overline{)42}$

11. $486 \div 18$

12. $16\overline{)336}$

13. $\begin{array}{r} 185 \\ - \ 78 \\ \hline \end{array}$

14. $\begin{array}{r} 748 \\ + \ 546 \\ \hline \end{array}$

15. $\begin{array}{r} 87 \\ * \ 12 \\ \hline \end{array}$

16. 496 / 8

17. $\begin{array}{r} 79 \\ * \ 57 \\ \hline \end{array}$

18. $\begin{array}{r} 9{,}017 \\ + \ 4{,}526 \\ \hline \end{array}$

19. Daniel planted 6 rows of beets in his garden, with 24 beets in each row. How many beets did he plant in all?

20. The tank of Mr. Washington's car holds about 16 gallons of gasoline. About how many gallons are in the tank when the gauge shows $\frac{1}{2}$ full?

21. When the gas tank is about $\frac{1}{4}$ full, Mr. Washington stops to fill the tank. If gasoline costs \$1.50 per gallon, about how much does it cost to fill the tank?

22. Joshua records his weight change every week. At the beginning of March, he weighed 87 pounds. His weekly weight fluctuations in March were +2, –1, +1 and +3 pounds. What was Joshua's total weight change for March? How much did he weigh at the end of the month?

Practice Set 45

Write all of your answers on a separate sheet of paper.

Write the letter of the picture that represents a fraction that is equivalent to the given fraction.

1. $\frac{1}{2}$

2. $\frac{6}{10}$

3. $\frac{4}{5}$

4. $\frac{2}{6}$

A.

X X X X X
X X X X X

B.

C.

D.

X X X
X X X

For each pair of fractions, write *yes* if the fractions are equivalent. Write *no* if the fractions are not equivalent.

5. $\frac{1}{6}, \frac{2}{12}$

6. $\frac{7}{12}, \frac{3}{4}$

7. $\frac{1}{2}, \frac{5}{10}$

8. $\frac{1}{3}, \frac{2}{9}$

9. $\frac{4}{5}, \frac{12}{20}$

10. $\frac{40}{50}, \frac{8}{10}$

11. $\frac{6}{8}, \frac{3}{4}$

12. $\frac{1}{4}, \frac{4}{12}$

Solve.

13. $1.20 * 5 = ■

14. 24 = 8 * ■

15. 9.2 + 6.3 = ■

16. 3 * ■ = $1.80

17. ■ ÷ 1,200 = 5

18. 14.0 − 7.4 = ■

19. 18 * ■ = 36

20. ■ ÷ 180 = 2

Write all of your answers on a separate sheet of paper.

Complete the frames-and-arrows problems.

21.

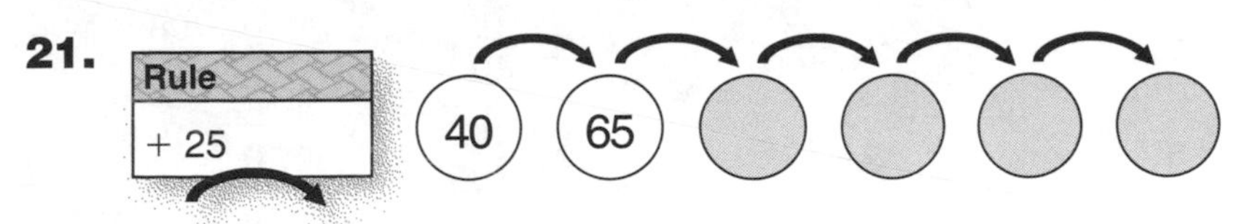

22.

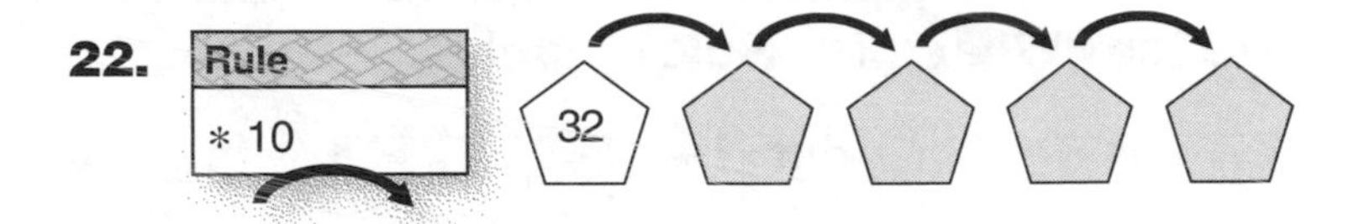

23.

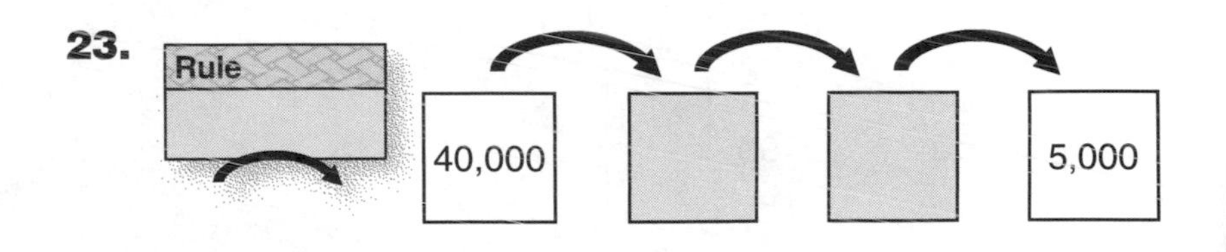

24.

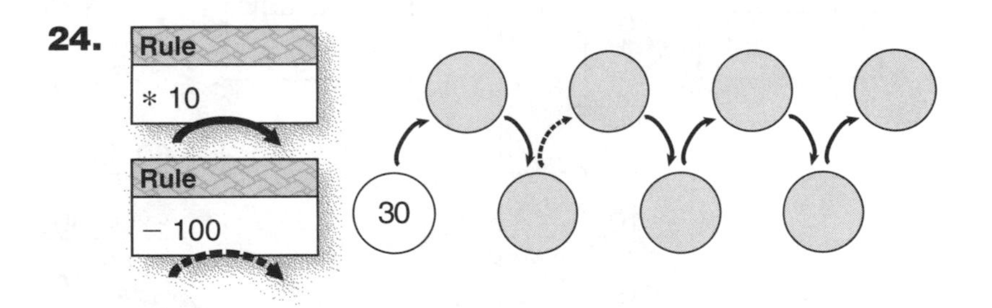

Practice Set 46

Write all of your answers on a separate sheet of paper.

Write an equivalent decimal for each fraction.

1. $\frac{1}{5}$ **2.** $\frac{74}{100}$ **3.** $\frac{3}{10}$ **4.** $\frac{98}{100}$

5. $\frac{4}{10}$ **6.** $\frac{4}{5}$ **7.** $\frac{39}{100}$ **8.** $\frac{19}{100}$

9. $\frac{7}{10}$ **10.** $\frac{55}{100}$ **11.** $\frac{2}{5}$ **12.** $\frac{81}{100}$

Complete the "What's My Rule?" tables.

13. Rule: out = in + 15

in	out
18	
34	
	56
48	
90	

14. Rule: out = in − 22

in	out
43	
34	
	9
18	
	77

15. Rule: out = in ∗ 4

in	out
0	
	8
4	
10	
15	

16. Rule: out = in / 3

in	out
18	
	12
60	
	30
	40

Practice Set 47

Write all of your answers on a separate sheet of paper.

Write <, >, or = to make each number sentence true.

1. $\frac{2}{5}$ ■ $\frac{7}{8}$

2. $\frac{1}{3}$ ■ $\frac{2}{3}$

3. $\frac{7}{8}$ ■ $\frac{1}{2}$

4. $\frac{5}{8}$ ■ $\frac{10}{16}$

5. $\frac{9}{10}$ ■ $\frac{2}{10}$

6. $\frac{6}{12}$ ■ $\frac{3}{6}$

7. $\frac{6}{30}$ ■ $\frac{3}{15}$

8. $\frac{9}{12}$ ■ $\frac{1}{5}$

Decide whether each fraction is less than $\frac{1}{2}$, equal to $\frac{1}{2}$, or greater than $\frac{1}{2}$. Write *less than $\frac{1}{2}$*, *greater than $\frac{1}{2}$*, or *equal to $\frac{1}{2}$*.

9. $\frac{3}{6}$

10. $\frac{10}{25}$

11. $\frac{12}{15}$

12. $\frac{10}{20}$

13. $\frac{7}{10}$

14. $\frac{1}{12}$

15. $\frac{9}{16}$

16. $\frac{12}{50}$

---

17. Write the number that has

4 in the tens place

7 in the hundred-thousands place

5 in the ones place

0 in the thousands place

6 in the hundreds place

8 in the ten-thousands place

Practice Set 47 (cont.)

Write all of your answers on a separate sheet of paper.

Solve.

18. $\begin{array}{r} 125 \\ +\ 76 \\ \hline \end{array}$

19. $\begin{array}{r} 43 \\ *\ 4 \\ \hline \end{array}$

20. $\begin{array}{r} 610 \\ *\ 8 \\ \hline \end{array}$

21. $\begin{array}{r} 810 \\ -\ 271 \\ \hline \end{array}$

22. $\begin{array}{r} 680 \\ -\ 596 \\ \hline \end{array}$

23. $\begin{array}{r} 47 \\ +\ 72 \\ \hline \end{array}$

24. $\begin{array}{r} 1{,}015 \\ -\ 450 \\ \hline \end{array}$

25. $\begin{array}{r} 600 \\ -\ 310 \\ \hline \end{array}$

26. $\begin{array}{r} 204 \\ +\ 329 \\ \hline \end{array}$

27. $\begin{array}{r} 8{,}134 \\ +\ 3{,}538 \\ \hline \end{array}$

28. $\begin{array}{r} 2{,}000 \\ -\ 199 \\ \hline \end{array}$

29. $\begin{array}{r} 460 \\ -\ 280 \\ \hline \end{array}$

30. It takes Sam about 35 minutes to get ready for school. If the bus comes by at 7:45 A.M., what time should Sam get up?

31. 6 * ■ = 18

32. 3 * 7 = ■

33. 16 / 4 = ■

34. 20 / ■ = 4

35. 24 / 8 = ■

36. 9 * 4 = ■

37. 8 * ■ = 64

38. 81 = ■ * 9

39. 2 * 30 = ■

40. 540 ÷ ■ = 90

Practice Set 48

Write all of your answers on a separate sheet of paper.

Complete.

1. If 9 counters are $\frac{1}{2}$, then ____ counters are the ONE.

2. If 4 counters are $\frac{1}{3}$, then ____ counters are the ONE.

3. If 7 counters are $\frac{1}{5}$, then ____ counters are the ONE.

4. If 10 counters are $\frac{2}{9}$, then ____ counters are the ONE.

5. If 6 counters are $\frac{2}{3}$, then ____ counters are the ONE.

6. If 15 counters are $\frac{5}{8}$, then ____ counters are the ONE.

Tell if each product is in the tens, hundreds, or thousands.

7. $5 * 47$

8. $22 * 95$

9. $6 * 850$

10. $3 * 21$

11. $16 * 16$

12. $27 * 75$

13. $4 * 125$

14. $82 * 59$

Write *true* or *false* for each number sentence.

15. $82 + 7 = 99$

16. $(5 * 2) + 6 = 60$

17. $(27 / 9) - 15 = 18$

18. $81 + 5 < 100$

19. $527 - 92 > 500$

20. $670 / 10 < 70$

Practice Set 49

Write all of your answers on a separate sheet of paper.

Use the spinner for items 1–5. Write *true* or *false* for each statement.

1. The chances of the paper clip landing on red are 3 out of 8.

2. The paper clip is 3 times as likely to land on red as on yellow.

3. The paper clip is least likely to land on blue.

4. The paper clip has the same chance of landing on blue as on yellow.

5. The paper clip has a $\frac{1}{6}$ chance of landing on green.

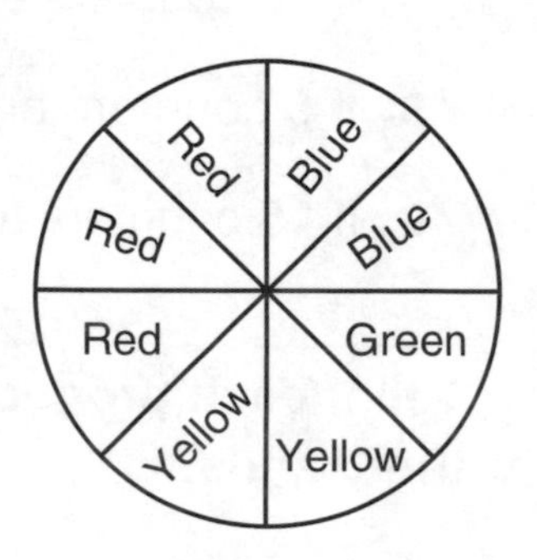

Write the digit in the hundredths place for each of the following.

6. 0.108 **7.** 13.313 **8.** 5.925 **9.** 4.078

Write 2 multiplication and 2 division facts for the following groups of numbers.

10. 4, 9, and 36 **11.** 6, 8, and 48

12. 10, 7, and 70 **13.** 7, 6, and 42

Solve.

14. 82 − 17 = ■ **15.** 70 + 26 = ■

16. 40 − 17 = ■ **17.** 16 − 8 = ■

18. 9 ∗ 9 = ■ **19.** 120 − ■ = 30

Practice Set 50

Write all of your answers on a separate sheet of paper.

Find the perimeter of each triangle. Convert measures of 12 inches or more to feet and inches.

1.

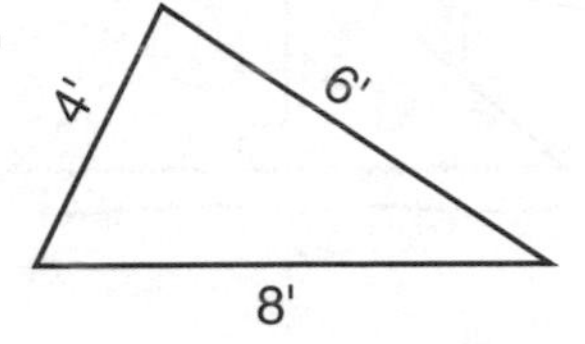

2.

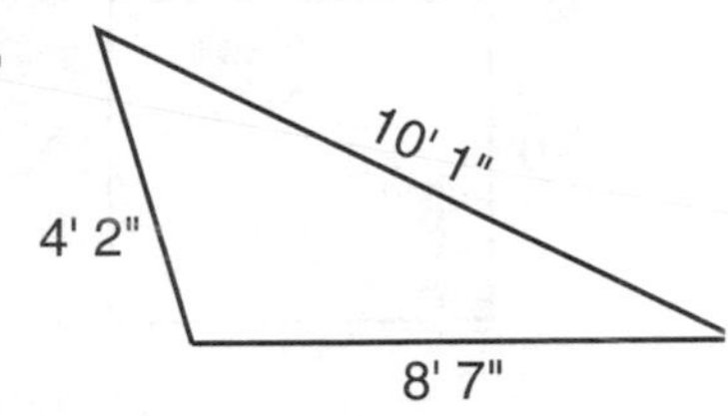

3.

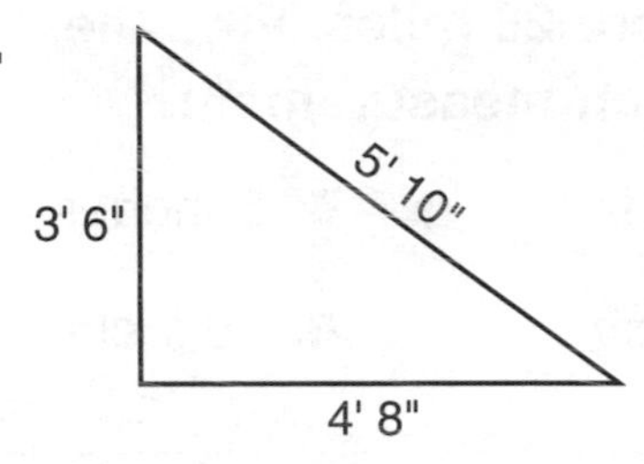

4.

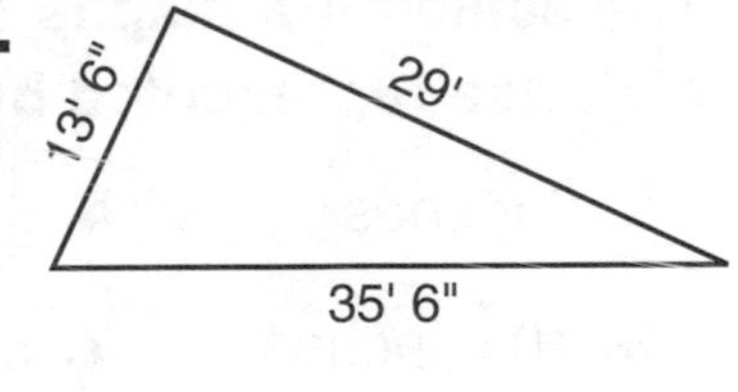

If 1 centimeter on a map represents 8 kilometers, write the distance represented by

5. 2 cm **6.** 3 cm **7.** 10 cm

8. 0.5 cm **9.** 2.5 cm **10.** 8.5 cm

---

11. Put these numbers in order from smallest to largest.
14,001 114,000 110.41 41,000

Write the next three numbers in each pattern.

12. 4, 8, 16 **13.** 85, 90, 95

14. 16, 12, 8 **15.** 2, 0, –2

Practice Set 51

Write all of your answers on a separate sheet of paper.

Find the perimeter of each figure.

1.

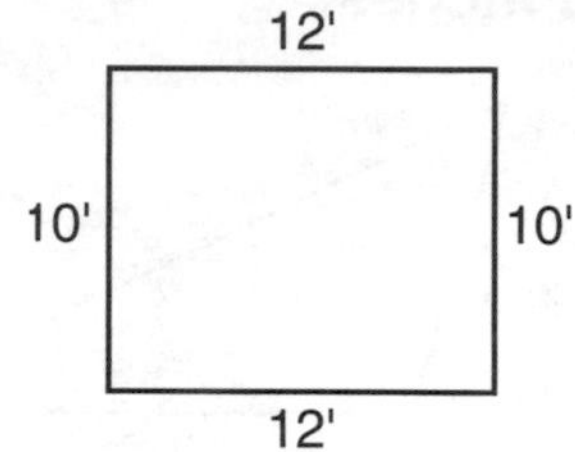

2.

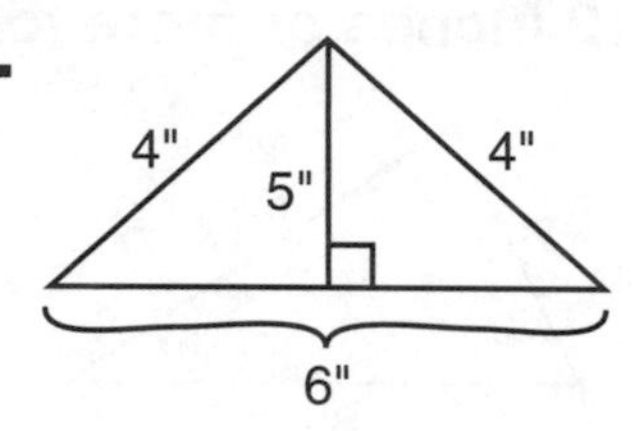

The scale for a map is 1 inch: 20 miles. Find the distance represented by each measurement.

3. 2 inches

4. $\frac{1}{2}$ inch

5. 5 inches

6. $10\frac{1}{2}$ inches

7. $8\frac{1}{4}$ inch

8. 16 inches

Solve.

9. $\begin{array}{r} 332 \\ -\ 140 \\ \hline \end{array}$

10. $\begin{array}{r} 38 \\ *\ 8 \\ \hline \end{array}$

11. $\begin{array}{r} 1{,}294 \\ +\ 5{,}729 \\ \hline \end{array}$

12. $\begin{array}{r} 600 \\ *\ 50 \\ \hline \end{array}$

13. $\begin{array}{r} 702 \\ 125 \\ +\ 311 \\ \hline \end{array}$

14. $\begin{array}{r} 39 \\ +\ 67 \\ \hline \end{array}$

15. $\begin{array}{r} 44 \\ +\ 35 \\ \hline \end{array}$

16. $\begin{array}{r} 92 \\ -\ 48 \\ \hline \end{array}$

17. $\begin{array}{r} 50 \\ -\ 16 \\ \hline \end{array}$

18. $\begin{array}{r} 87 \\ -\ 36 \\ \hline \end{array}$

19. $\begin{array}{r} 73 \\ -\ 58 \\ \hline \end{array}$

20. $\begin{array}{r} 509 \\ -\ 376 \\ \hline \end{array}$

Practice Set 52

Write all of your answers on a separate sheet of paper.

Find the area of each polygon in square units.

1.

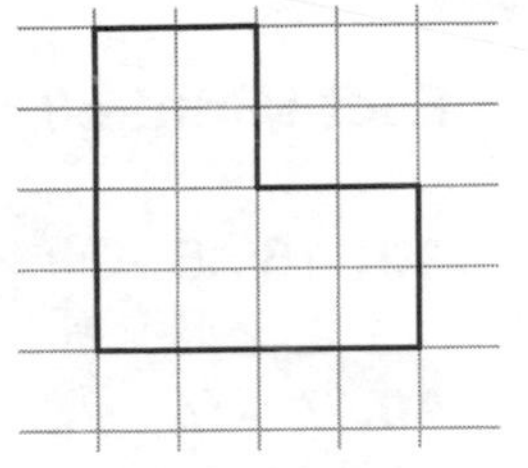

2.

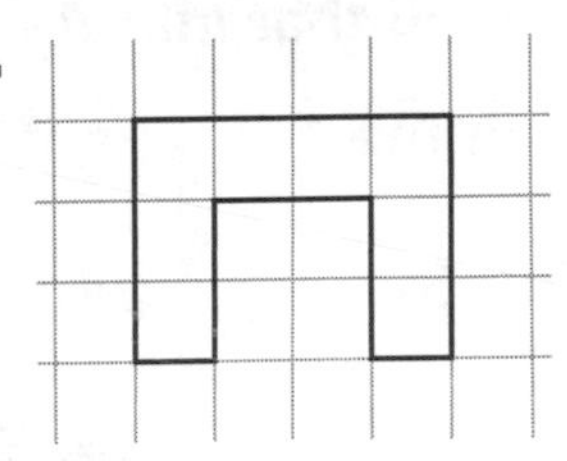

3.

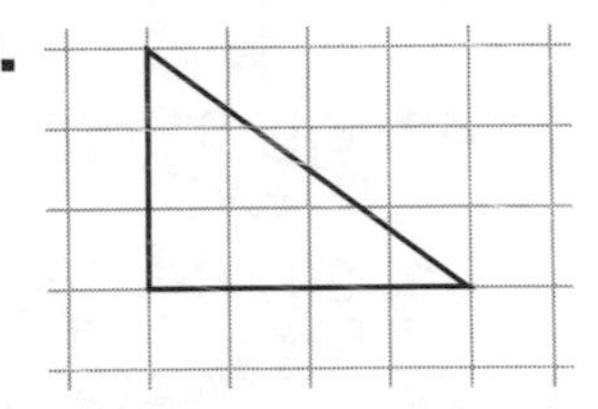

4.

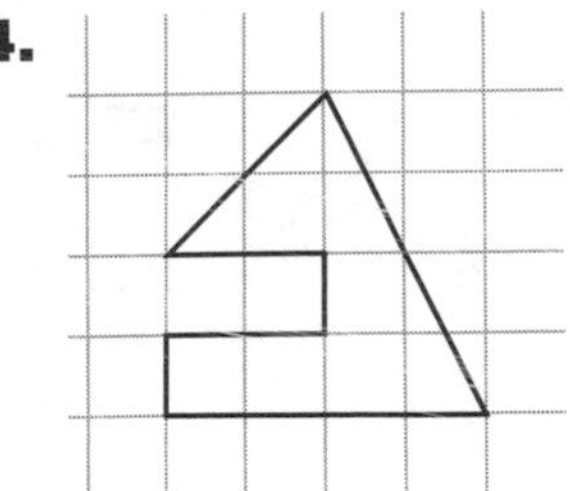

Use the following list of numbers to answer the questions.

18, 6, 7, 9, 11, 4, 14, 8, 11, 3, 6, 11

5. Which number is the smallest?

6. Which number is the largest?

7. What is the difference between the smallest and largest numbers?

8. Which number appears most often?

Practice Set 52 (cont.)

Write all of your answers on a separate sheet of paper.

For each Fact Minute below, do as many problems as you can in that minute.

Fact Minute 1

9. 9 * 6

10. 2 * 8

11. 4 * 5

12. 8 * 4

13. 3 * 8

14. 8 * 7

15. 49 / 7

16. 7 * 5

17. 8 / 4

18. 6 * 2

19. 45 / 9

20. 32 / 8

21. 2 * 5

22. 9 * 8

23. 6 * 3

Fact Minute 2

24. 4 * 8

25. 24 / 4

26. 5 * 9

27. 2 * 7

28. 3 * 4

29. 27 / 3

30. 6 * 8

31. 6 * 6

32. 2 * 6

33. 3 * 6

34. 7 * 3

35. 9 * 9

36. 63 / 9

37. 72 / 8

38. 81 / 9

Fact Minute 3

39. 18 / 6

40. 7 * 7

41. 54 / 9

42. 64 / 8

43. 20 / 5

44. 9 * 5

45. 7 * 4

46. 28 / 7

47. 48 / 8

48. 32 / 4

49. 9 * 7

50. 8 * 4

51. 9 * 5

52. 72 / 9

53. 48 / 6

Practice Set 53

Write all of your answers on a separate sheet of paper.

Martha raises vegetables, herbs, and flowers in her garden.

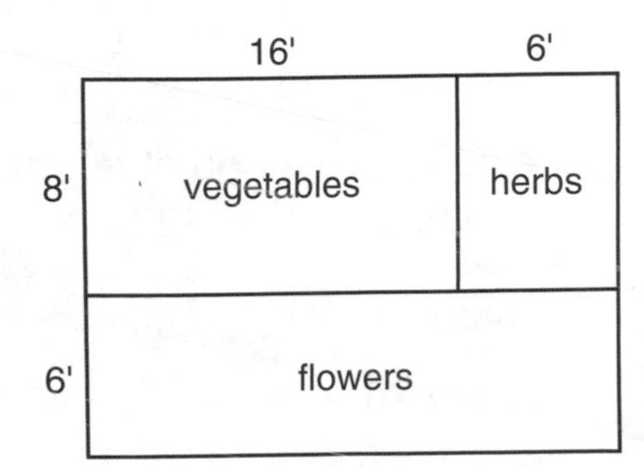

1. What is the total length of the garden? the total width of the garden?

2. What is the area of the vegetable section of the garden?

3. What is the area of the herb section of the garden?

4. What is the area of the flower section of the garden?

Rewrite the number sentences with parentheses to make them correct.

5. 7 * 9 − 4 = 59

6. 19 = 7 + 4 * 3

7. 31 − 14 − 5 = 12

8. 55 − 12 + 9 = 34

9. 4 * 9 + 4 * 12 = 84

10. 44 = 4 * 7 + 4

11. 9 * 1 + 7 * 8 = 576

12. 6 * 10 + 14 = 74

Practice Set 53 (cont.)

Write all of your answers on a separate sheet of paper.

Write the next three numbers in each pattern.

13. 36, 33, 30,

14. 10, 25, 40,

15. 48, 42, 36,

16. 140, 125, 110

---

17. Order these numbers from largest to smallest.
3,200 32,000 2,300 23,000

Solve.

18. ■ / 70 = 70

19. 6 * 30 = ■

20. 4,500 / ■ = 5

21. 9 * 90 = ■

22. 80 / 8 = ■

23. 30 * 80 = ■

24. ■ / 1,000 = 8

25. 1,400 / 700 = ■

26. 28 * ■ = 560

27. 800 = 8 * ■

28. 4 * 400 = ■

29. 30 * ■ = 1,500

Write numbers for the fractional parts shown in each picture.

30.

31.

32.

33.

Write as dollars and cents.

34. 18 dimes

35. 13 quarters

36. 35 nickels

37. 20 quarters and 6 dimes

38. Add the four amounts together.

Practice Set 54

Write all of your answers on a separate sheet of paper.

Find the area of each parallelogram.

1.

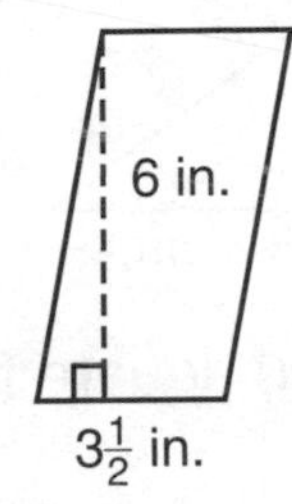

2.

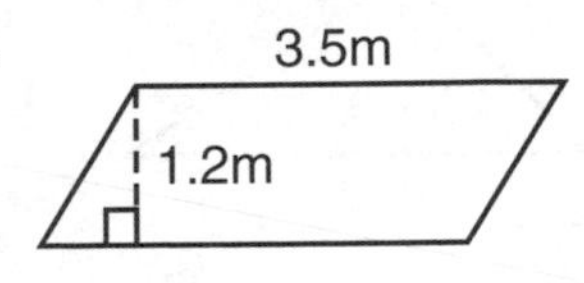

Write the number sentences with parentheses and solve.

3. Add 25 to the difference of 115 and 63.

4. Subtract the sum of 18 and 32 from 158.

5. Add 19 to the difference of 150 and 116.

6. Subtract the sum of 58 and 42 from 210.

Solve.

7. How many 25s in 300?

8. How many 50s in 1,200?

9. $8 \times 2{,}000$

10. $2{,}500 \times 3$

11. $1{,}500 \times 7$

12. $3{,}300 \times 30$

13. Without measuring, estimate this line segment to the nearest centimeter.

Practice Set 55

Write all of your answers on a separate sheet of paper.

Find the area or missing dimension of each triangle.

1.

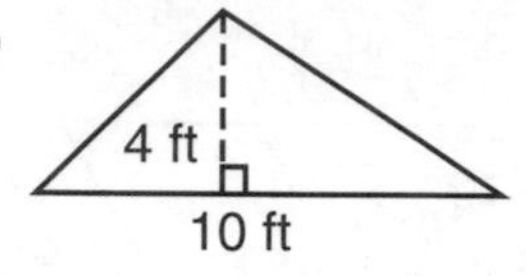

2.

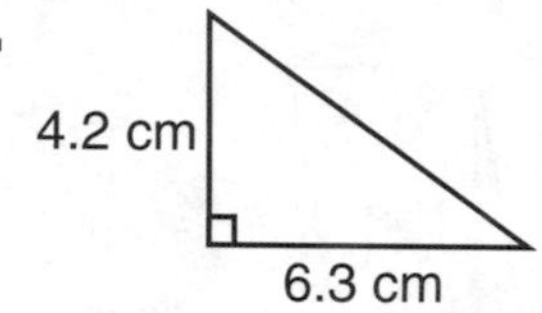

3. Area = 36 square meters

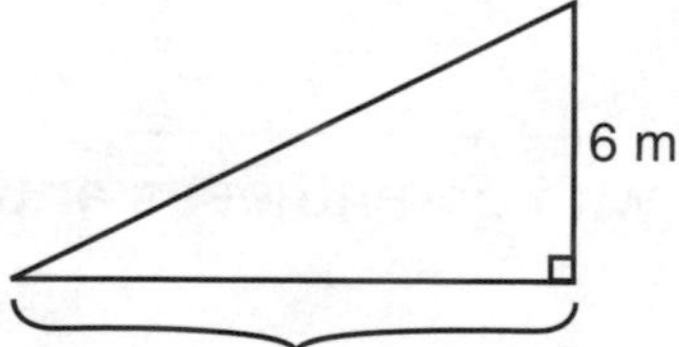

4. Area = 90 square feet

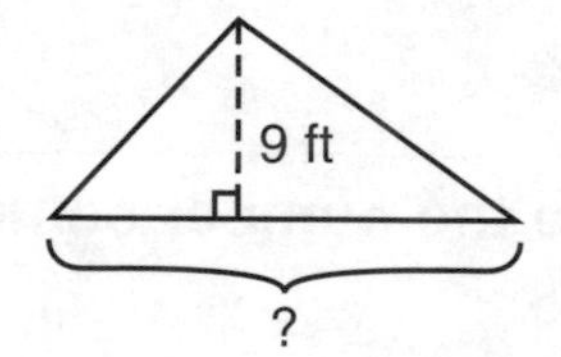

---

5. What time does the clock show? Write your answer to the nearest minute.

6. What time will it be in 50 minutes?

7. What time will it be in 128 minutes?

8. What time was it 2 hours and 25 minutes ago?

Practice Set 55 (cont.)

Write all of your answers on a separate sheet of paper.

Solve.

9. 3,389 + 1,974

10. 2,974 + 189

11. 26 * 4

12. 45 * 7

13. 40 * 500

14. 1.23 + 7.91

15. 4.6 + 4.9

16. 11.40 − 6.83

17. 12 * 9

18. 58 − 22

19. 205 + 832 + 117

20. 8,362 − 4,170

21. Use the clues to complete the place-value puzzle.

- Divide 72 by 9. Subtract 4 and write the result in the ones place.
- Double the number in the ones place. Write the result in the hundreds place.
- Multiply 8 * 10. Subtract 75. Write the result in the hundred-thousands place.
- Halve the number in the ones place. Multiply by 3 and write the result in the millions place.
- Divide 28 by the number in the ones place. Write the result in the ten-thousands place.
- Write the digit 1 in the remaining places.

1,000,000s	100,000s	10,000s	1,000s	100s	10s	1s

Practice Set 56

Write all of your answers on a separate sheet of paper.

Rename each decimal as a fraction with a denominator of 100 and as a percent.

1. 0.23 **2.** 0.52 **3.** 0.07 **4.** 0.10

Rename each percent as a fraction with a denominator of 100 and as a decimal.

5. 80% **6.** 15% **7.** 1% **8.** 24%

Write the value of the shaded part as a decimal, a fraction, and a percent.

9.

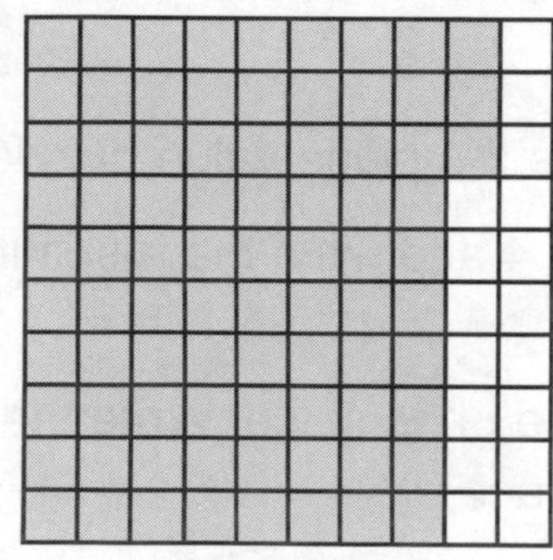

10.

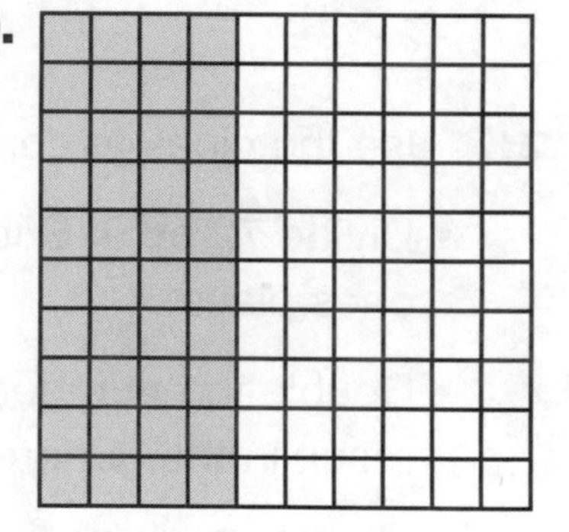

Write two equivalent fractions for the following numbers.

11. $\frac{1}{6}$ **12.** $\frac{3}{7}$

13. $\frac{3}{9}$ **14.** $\frac{6}{8}$

15. $\frac{11}{14}$ **16.** $\frac{3}{5}$

17. $\frac{4}{4}$ **18.** $\frac{12}{20}$

Practice Set 56 (cont.)

Write all of your answers on a separate sheet of paper.

For each Fact Minute below, do as many problems as you can in that minute.

Fact Minute 1

19. 13 − 4
20. 13 − 8
21. 12 − 7
22. 14 − 5
23. 17 − 9
24. 12 − 8
25. 12 − 4
26. 14 − 8
27. 12 − 3
28. 16 − 9
29. 15 − 8
30. 15 − 6
31. 10 − 3
32. 11 − 8
33. 9 − 5

Fact Minute 2

34. 4 − 0
35. 14 − 7
36. 9 − 9
37. 18 − 9
38. 10 − 1
39. 15 − 7
40. 13 − 7
41. 13 − 5
42. 10 − 6
43. 19 − 8
44. 12 − 5
45. 13 − 9
46. 13 − 6
47. 17 − 8
48. 11 − 7

Fact Minute 3

49. 12 − 9
50. 11 − 6
51. 10 − 7
52. 16 − 8
53. 11 − 4
54. 11 − 5
55. 10 − 2
56. 15 − 9
57. 12 − 6
58. 11 − 3
59. 14 − 9
60. 14 − 6
61. 16 − 7
62. 11 − 2
63. 10 − 4

Practice Set 57

Write all of your answers on a separate sheet of paper.

Match each decimal or fraction with the equivalent percent. Then write the letter that identifies the percent.

1. $\frac{2}{5}$	**A.** 50%
2. 0.60	**B.** 75%
3. $\frac{1}{2}$	**C.** 40%
4. 0.70	**D.** 80%
5. $\frac{3}{10}$	**E.** 10%
6. 0.80	**F.** 70%
7. 0.10	**G.** 60%
8. $\frac{3}{4}$	**H.** 30%

Complete the frames-and-arrows problems.

9.

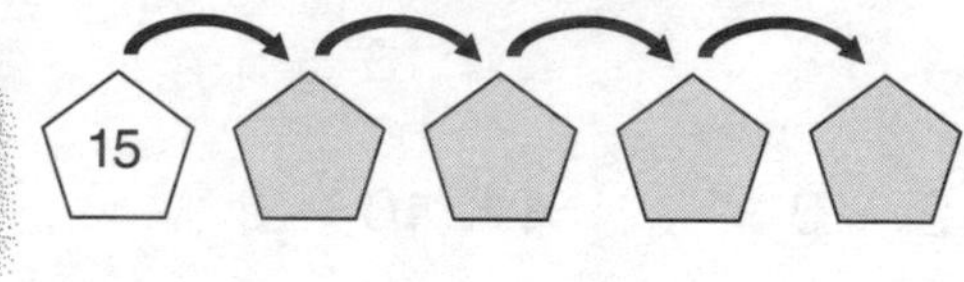

10.

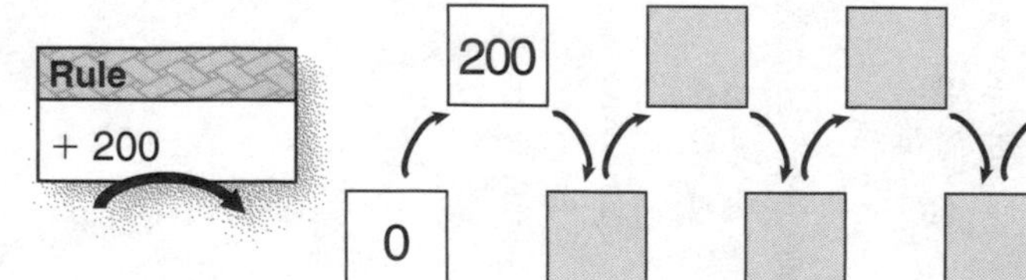

11.

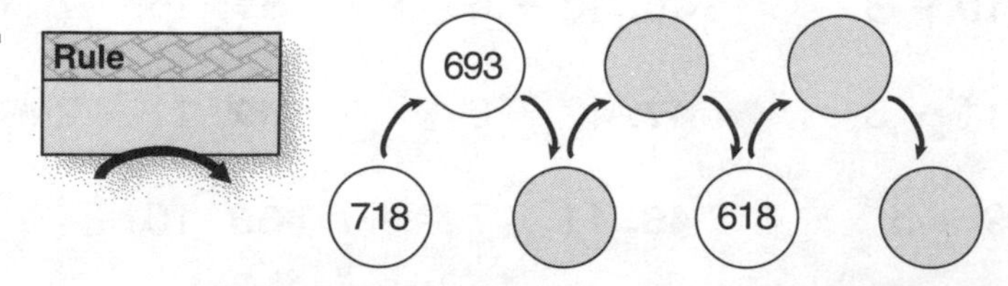

Practice Set 58

Write all of your answers on a separate sheet of paper.

There were 50 problems on a test.

1. Sara missed $\frac{1}{5}$ of the problems. She missed 0.20 of the problems. That's 20% of the problems.

a. How many problems did she miss?

b. $\frac{1}{5}$ of 50 = ____

c. 20% of 50 = ____

2. Erik missed $\frac{1}{10}$ of the problems. He missed 0.10 of the problems. That's 10% of the problems.

a. How many problems did he miss?

b. $\frac{1}{10}$ of 50 = ____

c. 10% of 50 = ____

Solve.

3. $40 - s = 35$

4. $55 = 18 + t$

5. $r + 9 = 51$

6. $90 - n = 43$

7. $76 + t = 206$

8. $12 = 357 - y$

9. $b - 50 = 750$

10. $543 + c = 812$

Solve these problems mentally.

11. 76,432 – 1,000

12. 76,432 – 100

13. 76,432 – 10,000

14. 76,432 – 10

---

15. Write the largest number you can using the following digits only once.

6 4 0 1 9 4 2 3 2

Practice Set 58 (cont.)

Write all of your answers on a separate sheet of paper.

Match the fraction with the shape that is the ONE for that fraction.

16. If 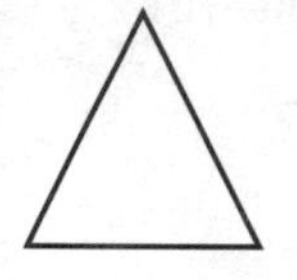 is $\frac{1}{3}$

A. 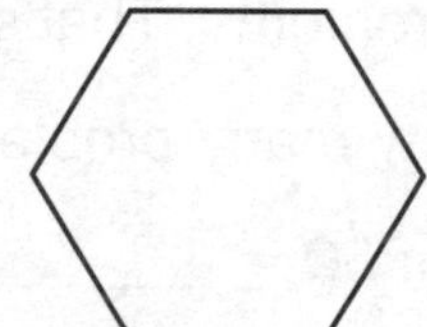

17. If 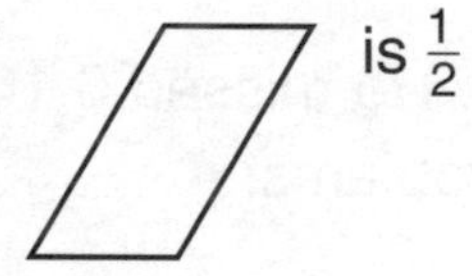 is $\frac{1}{2}$

B. 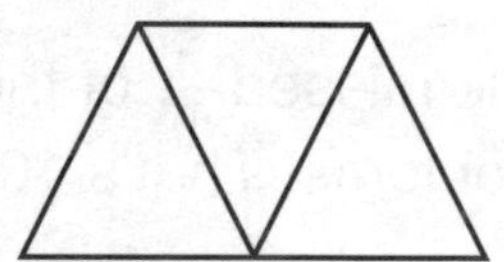

18. If 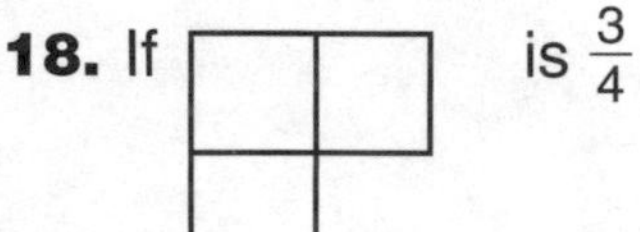 is $\frac{3}{4}$

C. 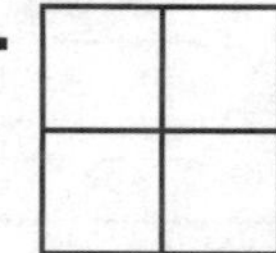

19. If 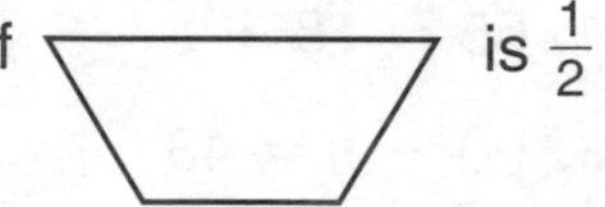 is $\frac{1}{2}$

D. 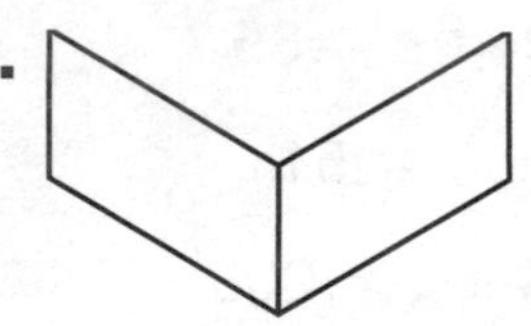

Complete.

20. 18 cm = ___ m

21. 3 m = ___ cm

22. ___ mm = 42 cm

23. ___ cm = 450 mm

24. 1.5 m = ___ cm

25. ___ mm = 8 cm

26. 100 mm = ___ cm

27. 200 cm = ___ m

28. ___ m = 75 cm

29. 155 cm = ___ mm

Practice Set 59

Write all of your answers on a separate sheet of paper.

Do not use a calculator to convert these fractions to percents.

1. $\frac{91}{100}$ **2.** $\frac{2}{10}$ **3.** $\frac{7}{20}$

4. $\frac{9}{25}$ **5.** $\frac{15}{50}$ **6.** $\frac{75}{125}$

Use a calculator to convert these fractions to percents.

7. $\frac{21}{30}$ **8.** $\frac{28}{70}$ **9.** $\frac{11}{88}$

10. $\frac{24}{80}$ **11.** $\frac{9}{36}$ **12.** $\frac{18}{48}$

Write the value of the digit 8 in the numerals below.

13. 589

14. 87,402

15. 719,538

16. 482,391

17. 8,946,326

Use digits to write the following numbers.

18. seventy-four million, nine thousand, sixty-four

19. nineteen and sixty-eight hundredths

20. four hundred nine and eight hundred twenty-seven thousandths

Practice Set 59 (cont.)

Write all of your answers on a separate sheet of paper.

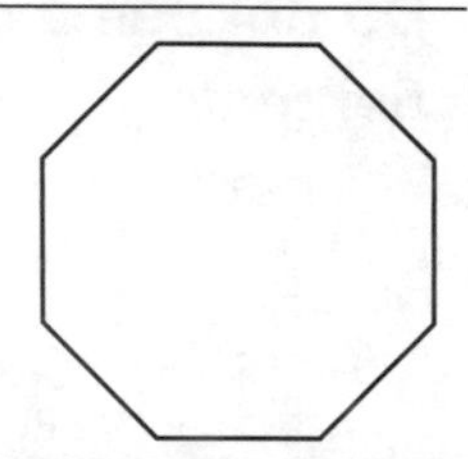

21. What kind of polygon is shown?

22. How many sides does it have?

23. If each side were 1.5 cm long, what would the perimeter be?

Complete the "What's My Rule?" tables.

24.

Rule
out = in + 38

in	out
9	
12	
15	
25	
100	

25.

Rule

in	out
7	490
10	700
	420
11	
3	210

26.

Rule
out = in − 108

in	out
80	
160	
	90
2,400	
	1,200

27.

Rule

in	out
160	4
440	11
	15
800	
30	$\frac{3}{4}$

Complete.

28. 140 minutes is the same as

■ hours and ■ minutes.

29. 63 hours is the same as

■ days and ■ hours.

 Use with or after Lesson 9.4.

Practice Set 60

Write all of your answers on a separate sheet of paper.

Ms. Juarez earns $35,000 a year. She made a budget to show how she spends her money. Write the percent of her budget that is

1. housing

2. clothing

3. food

4. transportation

5. savings

6. health

7. other

Mrs. Juarez's Budget	
Category	**Amount**
Housing	$15,000
Clothing	$2,500
Food	$3,750
Transportation	$5,200
Savings	$3,000
Health	$3,400
Other	$2,150

Solve.

8. $2.20 * 5 = ■

9. 72 = 9 * ■

10. 4.2 − 2.3 = ■

11. 5 * ■ = $3.50

12. ■ ÷ 1200 = 60

13. 28.0 − 6.4 = ■

14. 60 * ■ = 4200

15. ■ ÷ 36 = 20

Practice Set 61

Write all of your answers on a separate sheet of paper.

Match the expression with the number at the right. Write the letter of the number that corresponds to the expression.

1. 50% of 25,000	**A.** 600
2. 10% of 6,000	**B.** 5,000
3. 25% of 40,000	**C.** 12,500
4. 30% of 100,000	**D.** 10,000
5. 10% of 50,000	**E.** 30,000

Write the fractions in order from smallest to largest.

6. $\frac{1}{4}, \frac{1}{12}, \frac{1}{9}, \frac{1}{2}, \frac{1}{3}$

7. $\frac{4}{6}, \frac{4}{9}, \frac{4}{8}, \frac{4}{15}, \frac{4}{100}$

8. $\frac{15}{16}, \frac{5}{16}, \frac{3}{16}, \frac{1}{16}, \frac{7}{16}$

Rita checked the price of 1 pound of margarine at 5 different grocery stores. The prices she found were 85¢, 98¢, $1.09, 75¢, $1.08.

9. What is the maximum price?

10. What is the minimum price?

11. What is the range of prices?

12. What is the median price?

13. What is the mean (average) price?

Practice Set 61 (cont.)

Write all of your answers on a separate sheet of paper.

Complete.

14. $\frac{1}{4}$ of 40¢ is _____¢

15. $\frac{1}{2}$ of 50¢ is _____¢

16. $\frac{1}{3}$ of 60¢ is _____¢

17. $\frac{2}{3}$ of 30¢ is _____¢

Solve.

18. $16.25 + 5.13$

19. $7.4 + 5.13$

20. $41.02 + 6.89$

21. $5.97 + 3.6$

22. $6.6 - 2.1$

23. $15.43 - 3.06$

24. $12.98 - 7.8$

25. $59.35 - 11.63$

Write the amounts.

26. (Q) (Q) (Q) (D) (D) (D) (D) (N) (P) (P) (P) (P) (P)

27. [$1] [$1] (Q) (Q) (Q) (D) (D) (D) (D) (N) (P) (P)

28. [$5] [$1] [$1] [$1] [$1] (Q) (Q) (Q) (Q) (N) (N) (N)

29. [$100] [$100] [$20] [$5] [$1] [$1] [$1] (Q) (Q) (Q) (Q) (Q)

Practice Set 62

Write all of your answers on a separate sheet of paper.

Rewrite the product, correctly placing the decimal point.

1. 7 * 2.3 = 161

2. 0.9 * 42 = 378

3. 1.4 * 281 = 3,934

4. 0.03 * 510 = 153

5. 6.16 * 40 = 2,464

6. 2,435 * 7.2 = 17,532

Multiply.

7. 2.6 * 96

8. 0.51 * 43

9. 5.42 * 6

10. 8.7 * 12

11. 124 * 0.9

12. 1.64 * 52

In the numeral 28,490, the 8 stands for 8,000.

13. What does the 4 stand for?

14. What does the 2 stand for?

15. What does the 9 stand for?

16. What does the 0 stand for?

Tell whether each number sentence is *true* or *false*.

17. 14 + 8 = 22

18. 65 − 12 = 54

19. 18 = 39 − 14

20. 74 = 26 + 48

Practice Set 62 (cont.)

Write all of your answers on a separate sheet of paper.

Answer the questions by writing each fraction in simplest terms.

21. What fraction of the coins is pennies?

22. What fraction of the coins is nickels?

23. What fraction of the coins is dimes?

24. How much money is there in the whole group?

25. If you took away $\frac{1}{3}$ of the dimes, how much money would be left?

---

26. Use the clues to complete the place-value puzzle.

- Add 3 to the result of 71 – 68. Write the result in the hundredths place.
- Write the result of 54 / 9 in the ones place.
- Multiply 6 ∗ 12. Subtract 65. Write the result in the tens place.
- Double the number in the ones place. Then divide by 3 and write the result in the thousandths place.
- Divide 24 by 6. Add 5 and write the result in the tenths place.

10s	1s		0.1s	0.01s	0.001s
		.			

Practice Set 63

Write all of your answers on a separate sheet of paper.

Divide.

1. 28.4 / 4

2. 1.68 / 8

3. 211.5 / 9

4. 68.4 ÷ 6

5. 11.2 / 7

6. 36.5 ÷ 5

---

7. Without measuring, estimate the length of this line segment to the nearest inch

8. Make 100s.

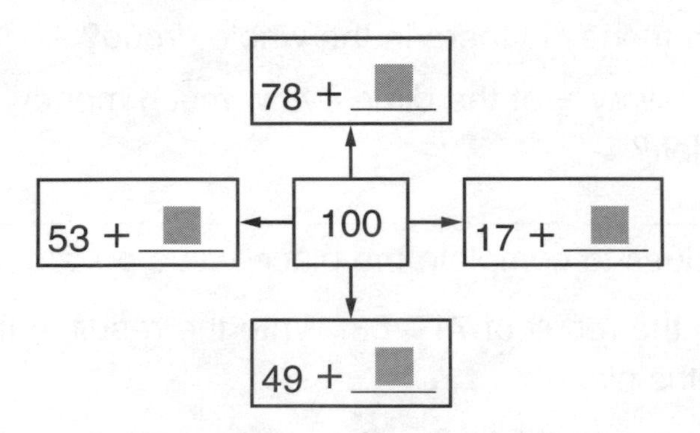

9. A square number is the product of a number multiplied by itself. For example, 25 is a square number because $5 * 5 = 25$. Which of the following numbers are square numbers?

16 28 36 100 54

Practice Set 64

Write all of your answers on a separate sheet of paper.

The following totals came up when Tina threw two dice.

6, 5, 3, 8, 2, 6, 9, 4, 11, 6, 6, 4, 8, 6

1. What is the maximum?

2. What is the minimum?

3. What is the range?

4. What is the median?

5. What is the mode?

6. What is the mean?

Fill in the missing numbers on the number lines.

7.

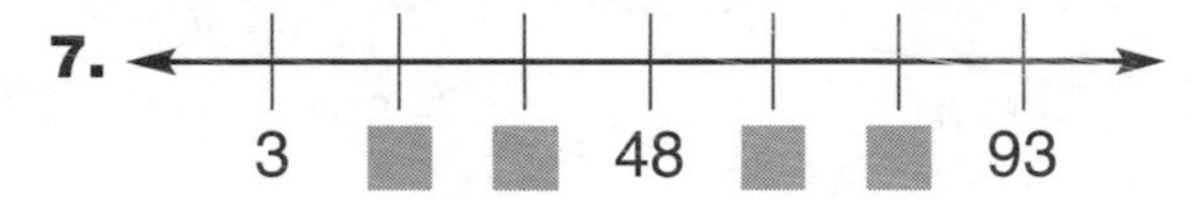

8.

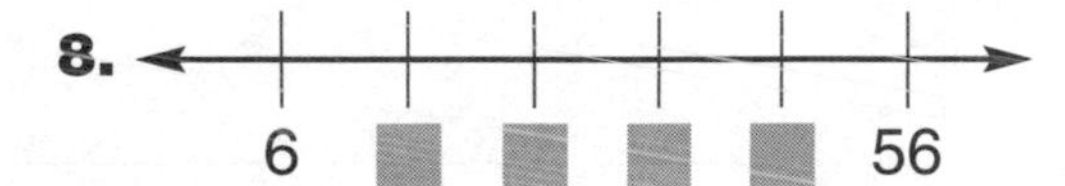

9.

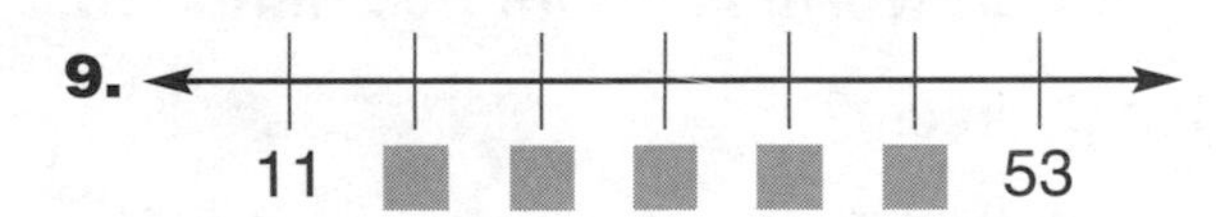

10.

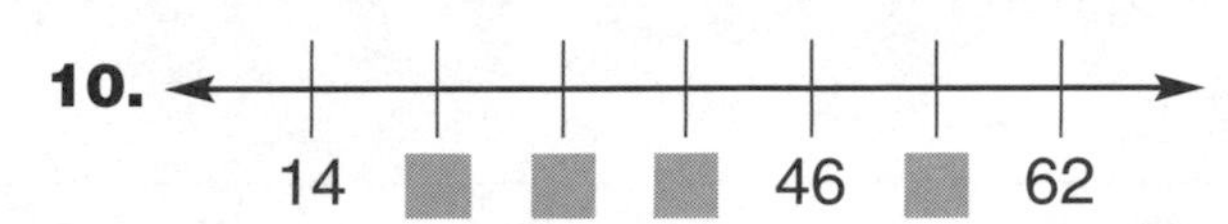

Measure the line segment to the nearest centimeter.

11. ____

12. ____________________________

Practice Set 65

Write all of your answers on a separate sheet of paper.

Write *yes* if the image is a reflection. Write *no* if the image is not a reflection.

1. Preimage | Image

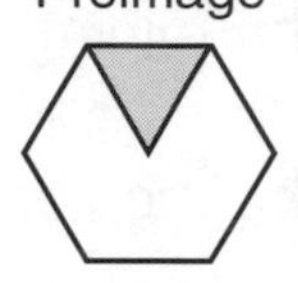

Line of Reflection

2. Preimage | Image

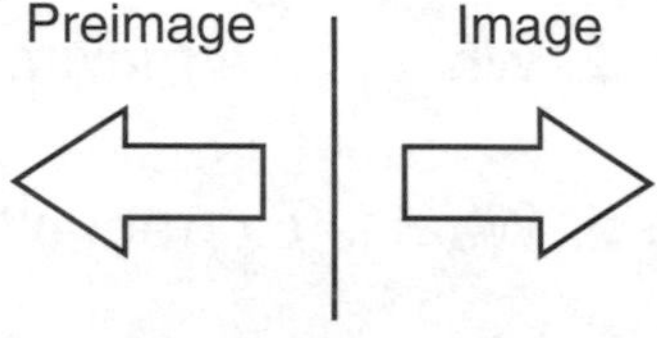

Line of Reflection

Complete.

3. 20 cm = ■ mm

4. 5,000 mm = ■ m

5. 20,000 mm = ■ cm

6. 2 m = ■ mm

7. 15 m = ■ cm

8. 2,000 mm = ■ cm

When straight, a threadworm is about 306 mm long.

9. What is its length in cm?

10. What is its length in m?

Suppose you spun a paper clip on the base of the spinner below 180 times.

11. How many times would you expect it to land on red?

12. How many times would you expect it to land on blue?

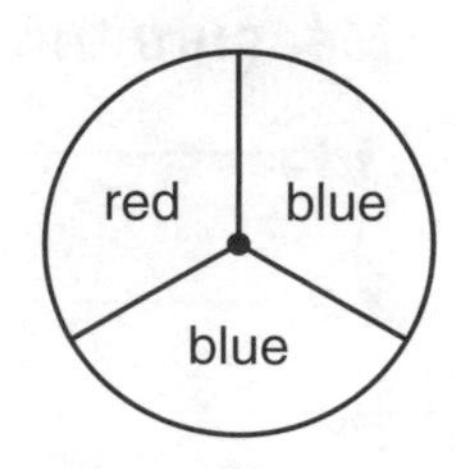

Practice Set 66

Write all of your answers on a separate sheet of paper.

Write *yes* if the image is a reflection. Write *no* if the image is not a reflection.

1. Preimage Image

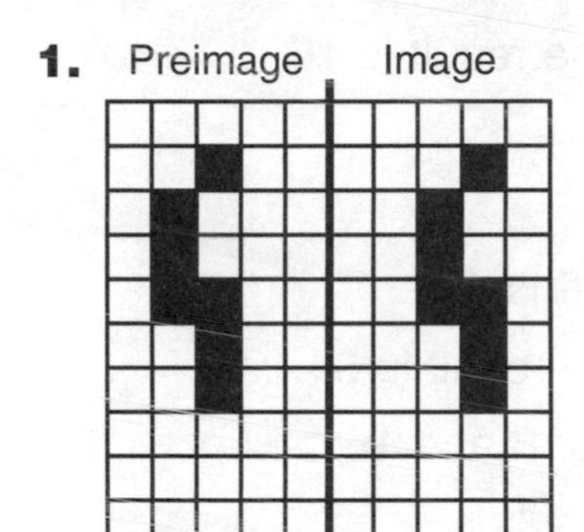

2. Preimage Image

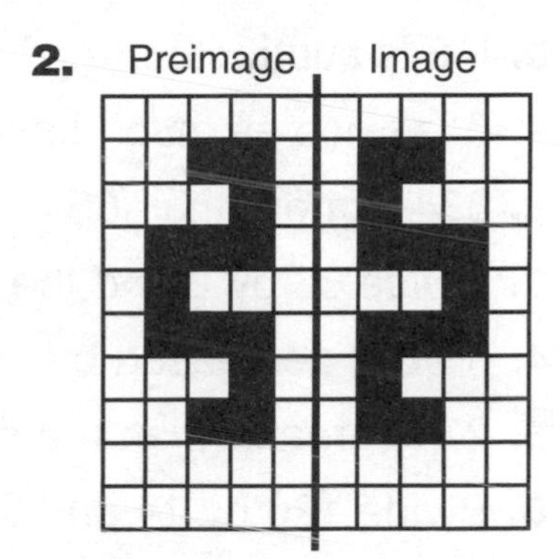

Write the following numbers with digits.

3. three hundred million, seventy-nine thousand

4. four billion, sixty-five million, seven hundred thousand

5. eighty-four billion, one hundred ninety-six million, forty thousand

---

Identify each angle. Write *acute, obtuse, straight,* or *reflex* for each angle.

6.

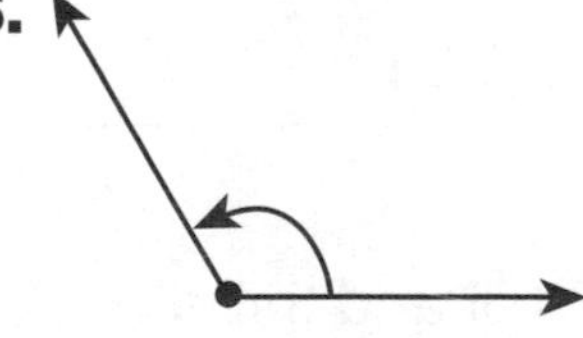

7.

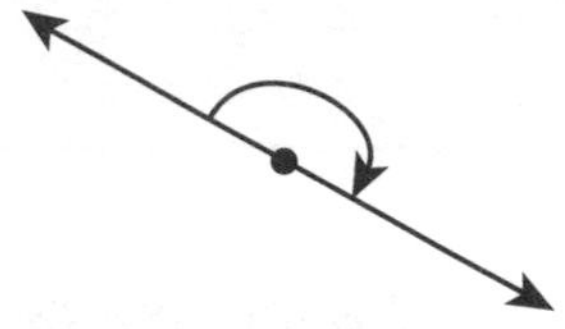

8.

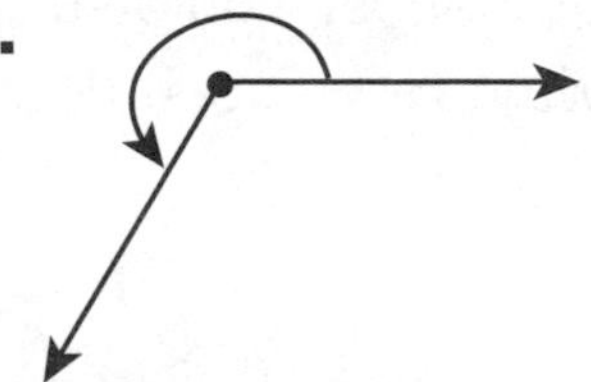

9.

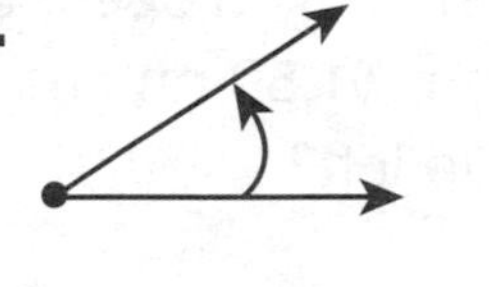

Practice Set 66 (cont.)

Write all of your answers on a separate sheet of paper.

Write number sentences for the following. Then tell whether they are *true* or *false*.

10. If 8 is subtracted from 24, the result is 16.

11. 6 is twice as much as 12.

12. 834 is more than 654.

13. Divide 86 by 2 and the result is 43.

14. If 98 is decreased by 16, the result is 84.

15. 27 is greater than the sum of 8 and 15.

16. Divide 126 by 14 and the result is 9.

17. 81 is the square number of 8.

Area = length *(l)* ∗ width *(w)*

Find the area in square units for each rectangle. Then write the number model.

18.

19. 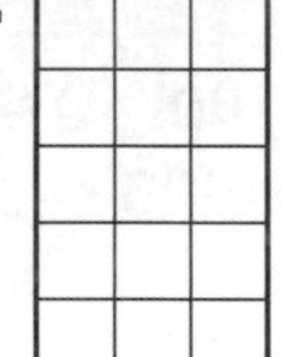

Solve.

20. You went to the mall with a $20 bill and three $1 bills. You spent $19.77 on groceries. You also spent $1.50 on bus fare each way. How much do you have left?

Practice Set 67

Write all of your answers on a separate sheet of paper.

Write *yes* if the figure has a vertical line of symmetry. Write *no* if it does not.

1. 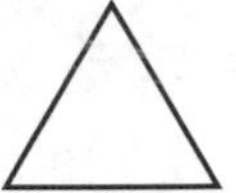

2. 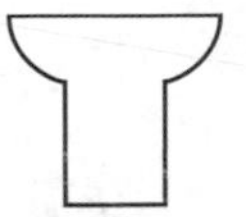

3. 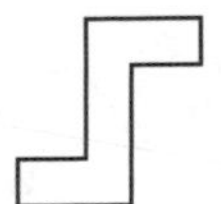

Write *yes* if the figure has a horizontal line of symmetry. Write *no* if it does not.

4. 

5. 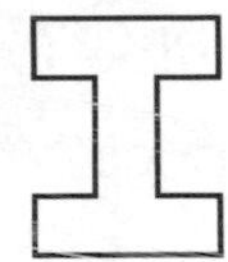

6. 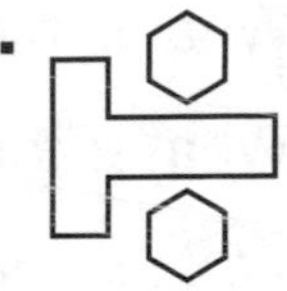

Rewrite the number sentences with parentheses to make them correct.

7. $9 * 12 - 3 = 81$

8. $9 * 12 - 3 = 105$

9. $15.8 = 2 * 6.5 + 2.8$

10. $18.6 = 2 * 6.5 + 2.8$

11. $7 * 1.1 + 4.2 * 12 = 58.1$

12. $5 * 12 + 2 - 4 = 58$

13. $5 * 12 + 2 - 4 = 66$

14. $8{,}140 = 110 * 50 + 24$

Complete.

15. 4 ft = ■ in.

16. 3 yd = ■ ft

17. 3 ft 5 in. = ■ in.

18. 2 yd 1 ft = ■ ft

19. 38 in. = ■ ft ■ in.

20. 9 ft = ■ yd

Practice Set 68

Write all of your answers on a separate sheet of paper.

Rename the following fractions as decimals.

1. $\frac{1}{10}$ **2.** $\frac{2}{4}$ **3.** $\frac{6}{16}$ **4.** $\frac{6}{10}$

5. $\frac{500}{1,000}$ **6.** $\frac{47}{100}$ **7.** $\frac{7}{8}$ **8.** $\frac{3}{4}$

9. $\frac{9}{16}$ **10.** $\frac{34}{100}$ **11.** $\frac{560}{1,000}$ **12.** $\frac{18}{18}$

Tim found 5 different prices for notebooks: 35¢, $1.15, $1.29, $2.18, $1.17.

13. What is the maximum price?

14. What is the minimum price?

15. What is the range of prices?

16. What is the median price?

17. What is the mean (average) price?

Fill in the missing numbers on the number lines.

18.

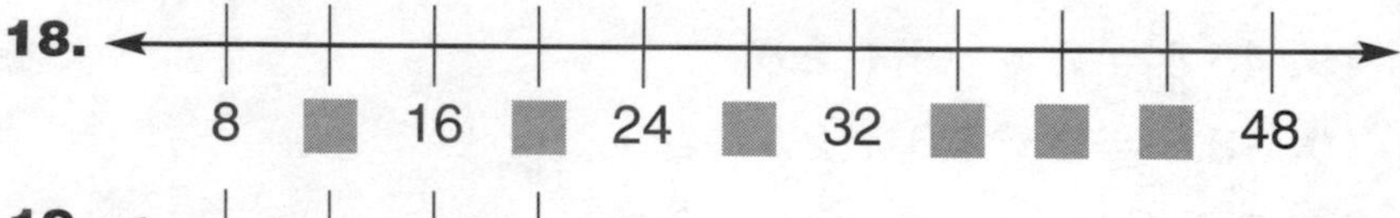

19.

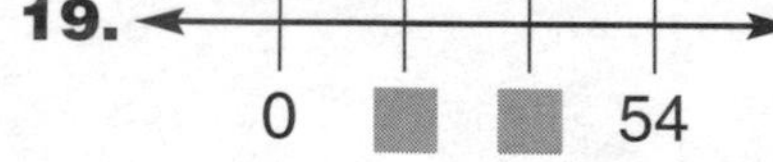

20.

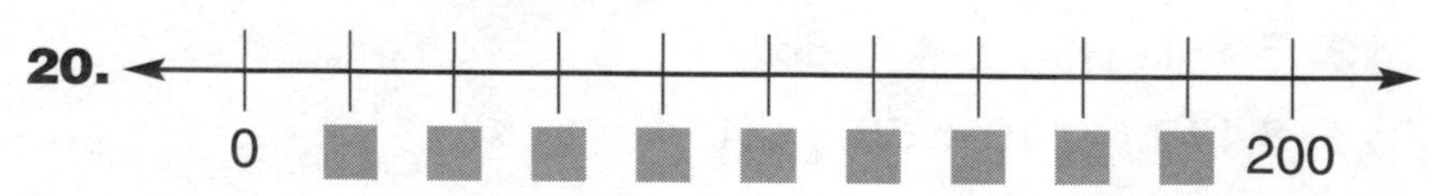

21.

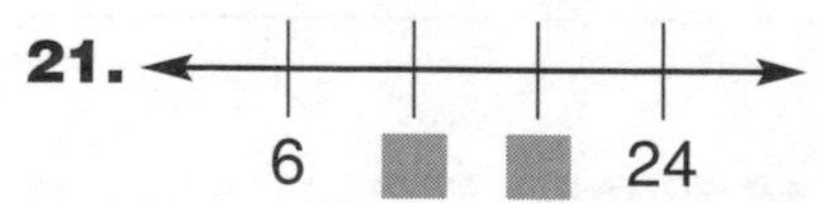

22.

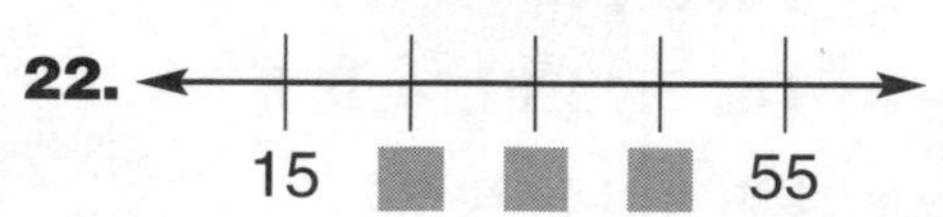

Practice Set 69

Write all of your answers on a separate sheet of paper.

1. What is the temperature difference, in °C, between Body Temperature and Room Temperature?

2. What is the temperature difference, in °F, between the boiling point and freezing point for water?

3. What is the temperature difference, in °F, between the freezing point for water and the freezing point for a salt solution? What is the difference in °C?

4. How much colder is –110°F than 7°F?

5. How much warmer is 42°C than –18°C?

6. Which is colder, –32°C or –32°F?

7. Which is warmer, 48°C or 108°F?

8. Imagine it is 22°C outside. Which would be a better activity: ice skating or bike riding?

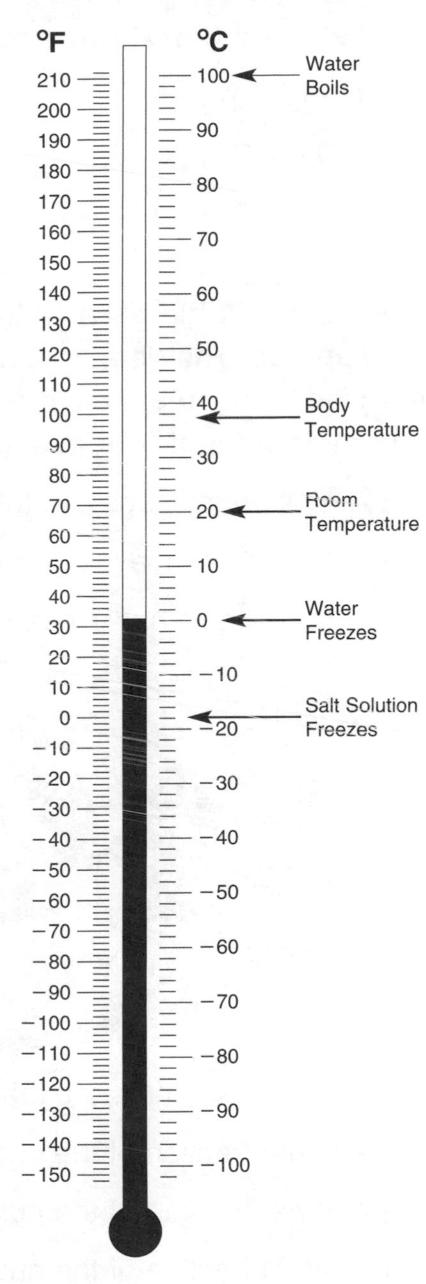

Write all of your answers on a separate sheet of paper.

What are the next three numbers in each pattern?

9. –15, –10, –5,

10. 0.04, 0.06, 0.08,

11. 0.44, 0.68, 0.92

Julie and Pattie have 18 bananas, 16 oranges, and 20 apples. They are making bags of mixed fruit, with 4 pieces of fruit in each bag. They can put any combination of fruit in each bag.

12. How many bags can they make?

13. How many pieces of fruit will they have left over?

14. If they also had 7 pears, how many bags could they make?

15. How many nuts are there?

16. What fraction of the nuts is peanuts?

17. What fraction of the nuts is acorns?

18. What fraction of the nuts is almonds?

Practice Set 70

Write all of your answers on a separate sheet of paper.

A nickel weighs about 5 grams. A liter of soda weighs about 1 kilogram.

Match the object with a possible weight. Write the letter of the possible weight.

1. a pair of scissors	**A.** 200 kg
2. a mug of hot chocolate	**B.** 1 kg
3. a loaf of bread	**C.** 1 g
4. a full-grown grizzly bear	**D.** 50 g
5. a safety pin	**E.** 350 g

Solve.

6. Two regular-size paper clips weigh about 1 gram. About how many paper clips would it take to weigh 10 grams?

7. About how many clips would it take to weigh 1 kilogram? (One kilogram = 1,000 grams.)

8. One ounce is about 30 grams. About how many regular-size paper clips are there in 1 ounce?

9. How many clips are there in 1 pound?

10. About how much does a box of 1,000 paper clips weigh if the empty box weighs 15 grams?

11. $8 * 9$	**12.** $96 / 8$
13. $60 \div 12$	**14.** $12 * 7$
15. $6 * 11$	**16.** $9 * 5$
17. $6 * 4$	**18.** $80 / 10$
19. $10 * 11$	**20.** $7 * 11$

Practice Set 70 (cont.)

Write all of your answers on a separate sheet of paper.

Copy and then complete the Powers of 10 Table.

The Powers of 10 Table

Millions	Hundred-Thousands	Ten-Thousands	Thousands	Hundreds	Tens	Ones
1,000,000				100		1
10 [100,000s]			10 [100s]			10 [0.1s]
		10*10*10*10				
	10^5		10^3			10^0

Practice Set 71

Write all of your answers on a separate sheet of paper.

Write the letter of the description that matches the polygon.

1. equilateral triangle

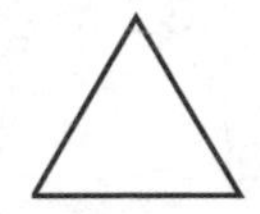

2. parallelogram

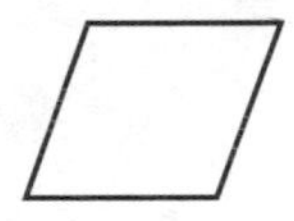

3. square

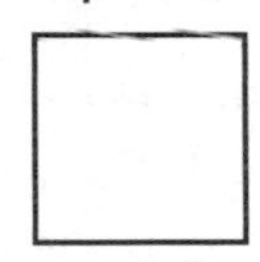

4. trapezoid

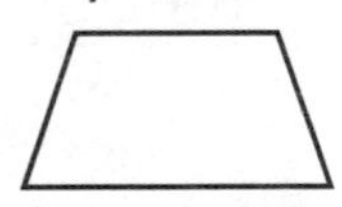

A. 4 sides of equal length, 4 right angles

B. no right angles, all sides the same length

C. only 1 pair of parallel sides

D. 2 pairs of parallel sides, no right angles

5. How much is $\frac{1}{8}$ of 32¢?

6. How much is $\frac{4}{9}$ of 54¢?

7. How much is $\frac{1}{10}$ of 80¢?

8. How much is $\frac{1}{3}$ of 90¢?

9. How much is $\frac{1}{5}$ of $2.20?

10. How much is $\frac{2}{3}$ of 27¢?

11. How much money, without tax, will I need for 3 audiotapes that cost $1.69 each?

Practice Set 72

Write all of your answers on a separate sheet of paper.

Find the area of each polygon.

1.

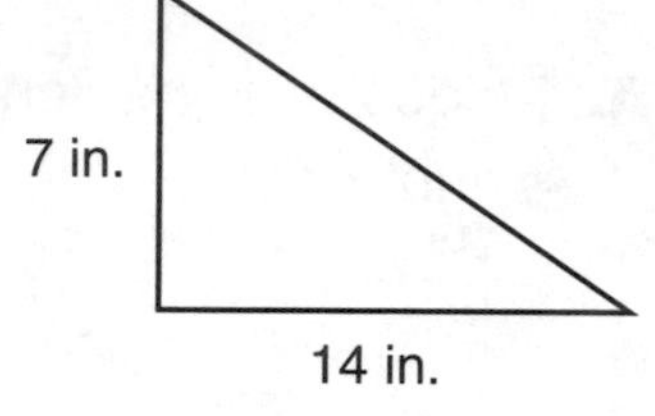

2.

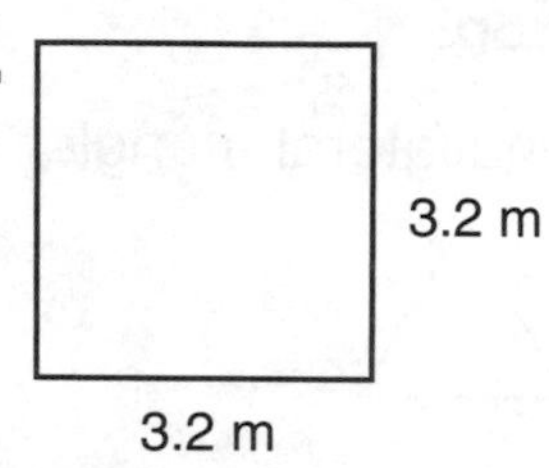

3.

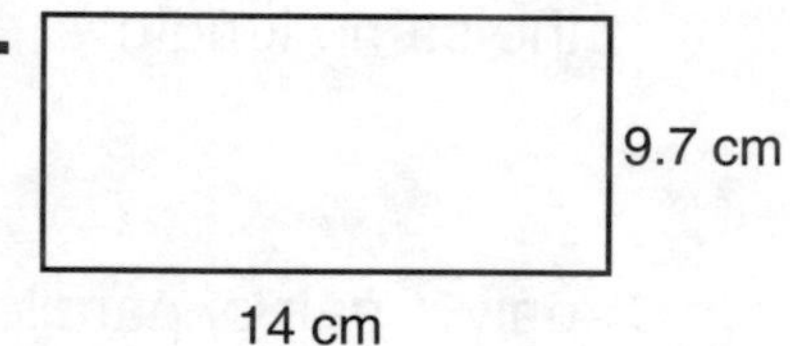

4.

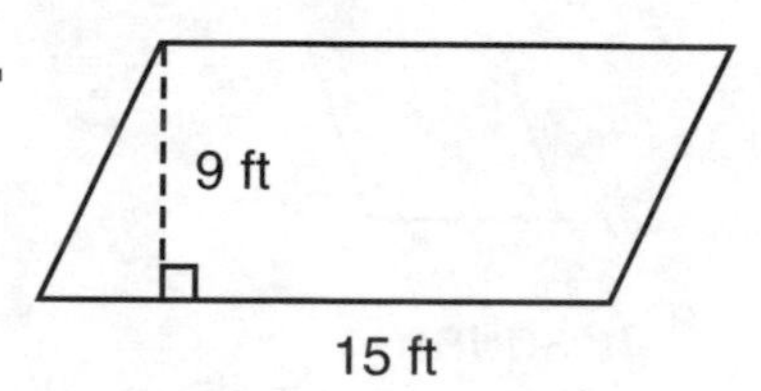

Solve.

5. $56 + 98$

6. $13 + 73$

7. $74 - 46$

8. $30 - 19$

9. $623 + 351 + 249$

10. $403 + 382$

11. $91 * 100$

12. $5,348 + 6,155$

13. $4,390 - 240$

14. (80 + 80) * 5

15. 36 + (7 * 9)

16. 60 + 450 + 338

17. (40 * 5) − 30

Practice Set 72 (cont.)

Write all of your answers on a separate sheet of paper.

In each set of problems below, do as many exercises as you can in one minute. Ask someone to time you.

Problem Set 1	Problem Set 2	Problem Set 3
18. 14 − 6	**33.** 84 / 7	**48.** 72 / 6
19. 7 / 7	**34.** 12 * 11	**49.** 44 / 4
20. 96 / 8	**35.** 7 + 18	**50.** 121 − 11
21. 4 * 111	**36.** 3 * 8	**51.** 9 * 9
22. 11 − 4	**37.** 4 * 10	**52.** 144 / 12
23. 12 + 5	**38.** 11 * 4	**53.** 7 * 8
24. 14 + 8	**39.** 3 * 11	**54.** 4 * 70
25. 12 − 7	**40.** 12 + 9	**55.** 144 / 12
26. 54 / 9	**41.** 4 * 7	**56.** 21 − 3
27. 9 * 12	**42.** 11 + 11	**57.** 55 + 5
28. 14 / 7	**43.** 81 / 9	**58.** 6 * 7
29. 24 / 3	**44.** 54 / 6	**59.** 9 * 9
30. 5 + 6	**45.** 50 + 60	**60.** 500 + 600
31. 18 / 2	**46.** 40 / 10	**61.** 32 / 8
32. 4 * 80	**47.** 27 / 3	**62.** 36 / 9

Practice Set 73

Write all of your answers on a separate sheet of paper.

Find the volume of each rectangular prism.

Volume = length × width × height
= area of base × height

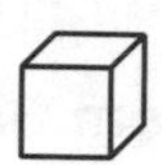

1 cubic unit

1.

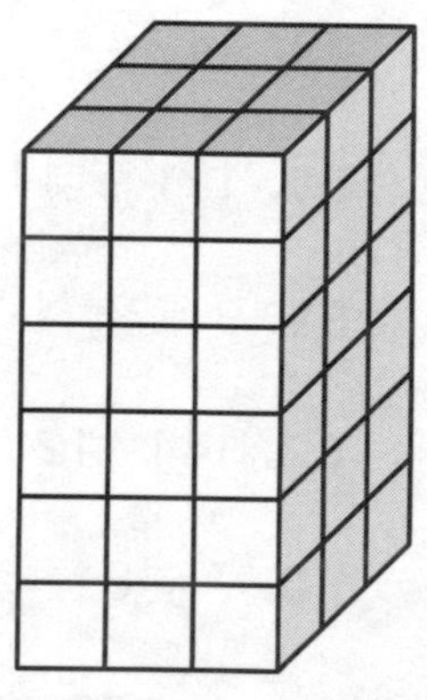

Volume = ■ cubic units

2.

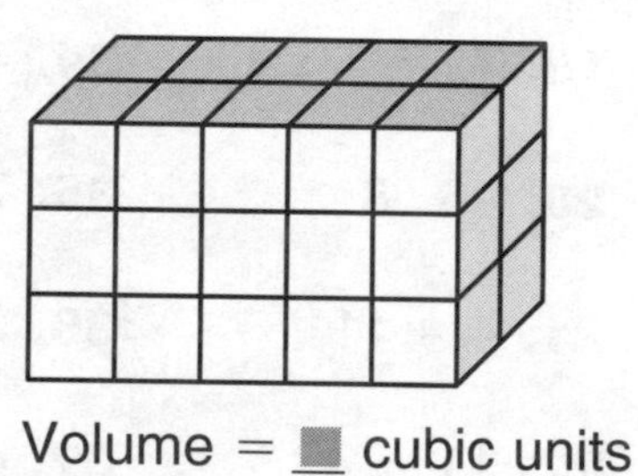

Volume = ■ cubic units

3.

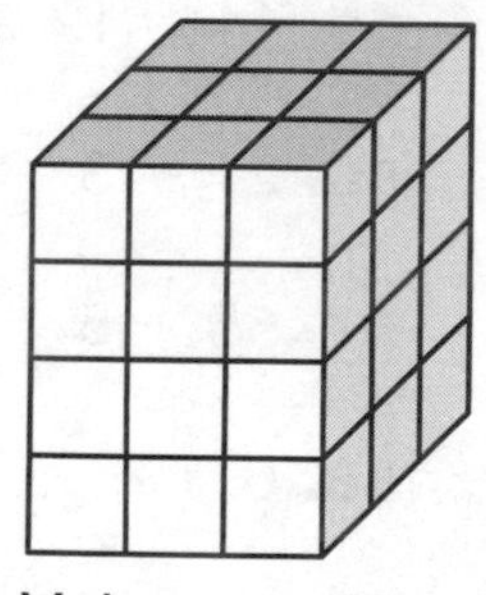

Volume = ■ cubic units

4.

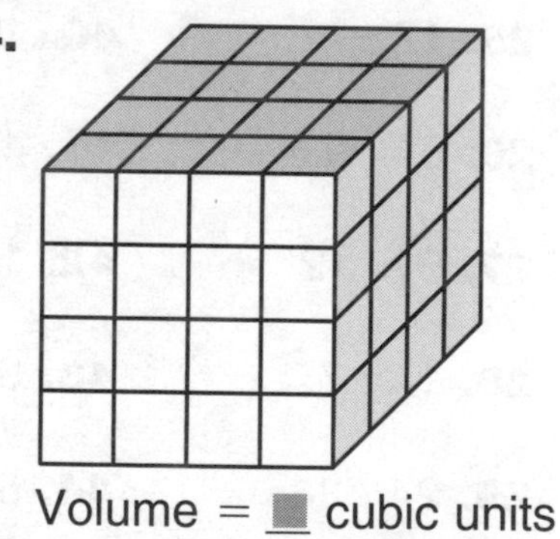

Volume = ■ cubic units

Write two equivalent fractions for each of the following numbers.

5. $\frac{1}{3}$ **6.** $\frac{3}{4}$ **7.** $\frac{3}{6}$ **8.** $\frac{5}{12}$

9. $\frac{10}{16}$ **10.** $\frac{14}{7}$ **11.** 1 **12.** $\frac{6}{9}$

Write all of your answers on a separate sheet of paper.

A super-sized pizza is divided into 12 pieces.

13. John ate 3 pieces of the pizza and Aaron ate 2 pieces. What fraction of the pizza was left?

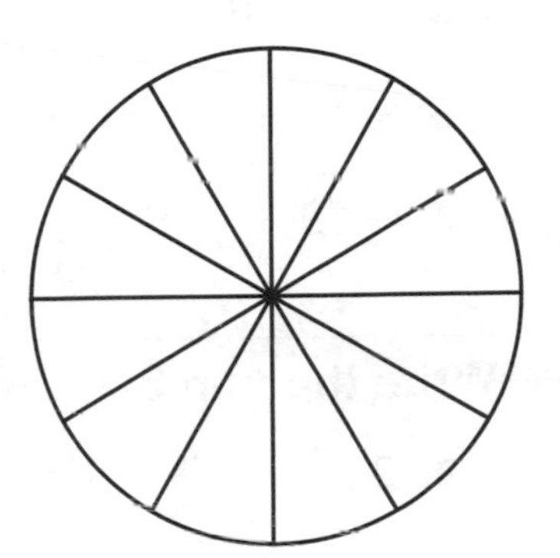

14. Charlie, Loren, and Travis each ate 1 piece of pizza. What fraction of the pizza did they eat?

15. The next day Mrs. Murphy took 2 pieces of pizza for lunch. What fraction of the pizza did she take?

Complete the missing factors.

16. 70 * ■ = 2,100

17. ■ * 4 = 360

18. ■ * 80 = 6,400

19. 12 * ■ = 960

20. 40 * ■ = 480

21. ■ * 50 = 3,500

22. 6 * ■ = 360

23. ■ * 7 = 840

Estimate the total cost.

24. 2 tape dispensers that cost $4.65 each

25. 12 magazines that cost $2.99 each

26. 9 scissors that cost 45¢ each

Practice Set 74

Write all of your answers on a separate sheet of paper.

Add or subtract.

1. $6 - (-10)$

2. $5 + (-1)$

3. $-6 + 6$

4. $17 - (-2)$

5. $-11 - (-15)$

6. $16 - (-7)$

7. $-19 + (-6)$

8. $-21 - 9$

Write the numbers from least to greatest.

9. $2.3, -7, \frac{4}{9}, -1.5, 8.3, -0.2$

10. $-11, 3\frac{2}{5}, 1.85, -5.5, 4, -\frac{8}{10}$

11. $\frac{17}{10}, -4, 9.9, 9.09, -3.7, 1.07$

Solve.

12. 540 / 6 = ■

13. 250 * 80 = ■

14. 640 = 8 * ■

15. 20 * 300 = ■

16. 60 * ■ = 2,400

17. ■ ÷ 50 = 6

18. 5,600 / 700 = ■

19. 360 * ■ = 7,200

20. ■ ÷ 5 = 35

21. 9 * 200 = ■

22. 540 ÷ ■ = 90

23. 110 * 120 = ■

Practice Set 74 (cont.)

Write all of your answers on a separate sheet of paper.

Rename the following numbers as percents.

24. $\frac{1}{4}$ **25.** 0.75 **26.** 1.00 **27.** $\frac{57}{100}$

28. $\frac{3}{20}$ **29.** $\frac{10}{25}$ **30.** 0.4 **31.** $\frac{37.5}{100}$

32. $\frac{4}{5}$ **33.** 1.125 **34.** $\frac{765}{1,000}$ **35.** $\frac{6}{15}$

---

1 km = 1000 m; 1 m = 100 cm
1 cm = 10 mm

Complete.

36. 2 km = ■ cm

37. 25,000 mm = ■ m

38. 1,800 m = ■ km

39. 30 km = ■ m

40. 3.3 cm = ■ mm

41. 670 cm = ■ mm

Who Am I?

42. Clue 1: I am a whole number less than 5.

Clue 2: If you multiply me by 3, the result is more than 10.

43. Clue 1: I am a negative number greater than -8.

Clue 2: If you add me to 7, the sum is 2.

44. Clue 1: I am a fraction between $\frac{1}{2}$ and 1.

Clue 2: My denominator is 5.

Clue 3: If you add me to $\frac{1}{4}$, the sum is greater than 1.

Practice Set 75

Write all of your answers on a separate sheet of paper.

Complete.

1. 4 cups = ___ pints

2. 3 quarts = ___ cups

3. ___ gallons = 8 quarts

4. ___ pints = 10 cups

5. $1\frac{1}{2}$ quarts = ___ cups

6. $3\frac{1}{2}$ gallons = ___ quarts

---

7. What time does the clock show? Write your answer to the nearest minute.

8. What time will it be in 35 minutes?

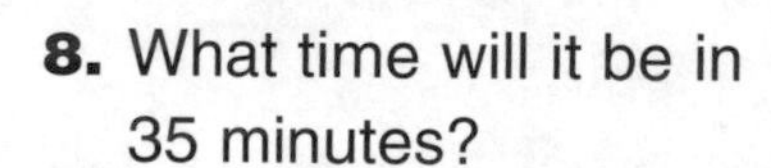

9. What time will it be in 88 minutes?

Find the percent of the following.

10. 70% of 10

11. 25% of 80

12. 75% of 12

13. 50% of 64

14. 24% of 25

15. 150% of 22

16. 80% of 50

17. 90% of 100

18. 33% of 1,000

19. 12% of 200

20. 6% of 50

21. 15% of 20

Solve these problems mentally.

22. 934,167 – 1,000 = ■

23. 934,167 – 100 = ■

24. 934,167 – 10,000 = ■

25. 934,167 – 10 = ■

Practice Set 76

Write all of your answers on a separate sheet of paper.

Write the missing numbers.

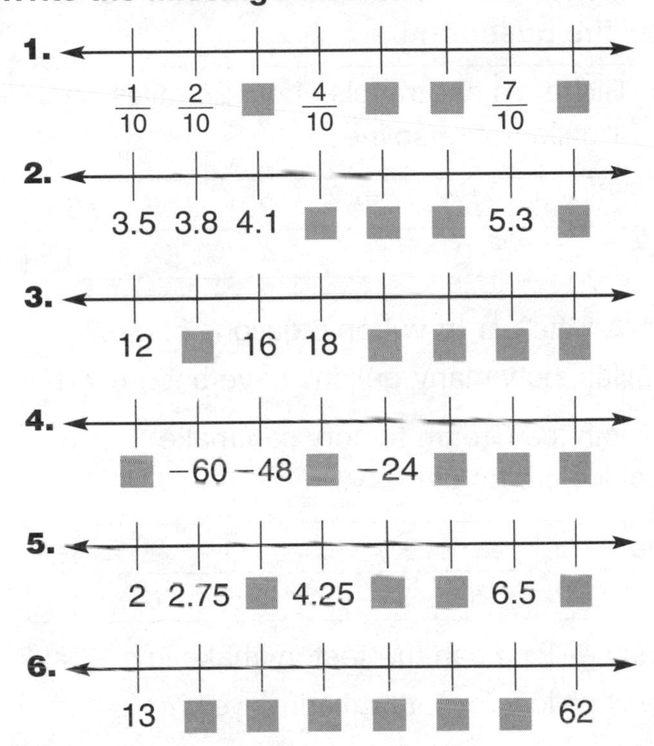

Solve.

7. Which temperature is colder, –17°C or –32°C?

8. Which is colder, +20°C or –13°C?

9. You have 62 cookies. You give 4 cookies to each of your friends until you run out of cookies. How many friends received cookies?

10. You have 62 cookies and you want to share them equally among you and 17 friends. How many will each person get?

Practice Set 77

Write all of your answers on a separate sheet of paper.

Write your own rate tables for the problems below. Then answer the questions.

Example Richard's car travels about 25 miles on 1 gallon of gasoline.

miles	25	50	75	100	125	150
gallons	1	2	3	4	5	6

1. How far can the car travel on 6 gallons of gas?

2. At 125 miles, how many gallons have been used?

The Sweet Tooth Ice Cream factory can make 100 gallons of ice cream per day.

gallons	100	■	■	■	■	■	■
day	1	2	3	4	5	6	7

3. How many gallons can the factory make in a week?

4. How many gallons can it make in a year?

Rita is a seamstress. She can make 3 dresses in 2 hours.

dresses	3	■	■	■	■	■
hours	2	4	6	8	10	12

5. How many dresses can Rita make in an 8-hour day?

6. If she takes off an hour for lunch, how many dresses will she make?

Write <, >, or = to make a true number sentence.

7. 29 + 33 ■ 45 + 17

8. 20 + 30 + 40 ■ 35 + 55

9. 35 + 13 ■ 32 + 18

10. 130 − 18 ■ 55 * 2

11. 13 + 12 ■ 48 ÷ 2

12. 118 + 220 ■ 1,000 − 692

Practice Set 77 (cont.)

Write all of your answers on a separate sheet of paper.

Find the percent of the following.

13. 34% of 50 **14.** 25% of 60 **15.** 15% of 100

16. 60% of 600 **17.** 1% of 20 **18.** 10% of 77

19. 5% of 20 **20.** 66% of 1,000 **21.** 22% of 150

22. 45% of 900 **23.** 10% of 300 **24.** 100% of 93

Solve each number sentence by finding the value of the variable.

25. $A = (8 * 9) \div 3$

26. $64 / B = 32 \div 2$

27. $C = (23 + 9) \div 8$

28. $45 + (11 * 6) = D$

29. $43 - E = 80 / 2$

30. $F / 19 = 570$

31. $(60 - 18) \div 6 = G$

32. $(13 * 20) - H = 10^2$

Complete.

33. $10^4 =$ ■

34. $10^{■} = 100,000$

35. $600 = 6 * 10^{■}$

36. $10 * 10 * 10 * 10 = 10^{■}$

37. 10 to the eighth power = ■

38. $3.0 * 10^6 =$ ■

39. 10 to the ■ power = 1,000

Practice Set 78

Write all of your answers on a separate sheet of paper.

Write your own rate tables for the problems below. Then answer the questions.

It takes Christie 3 minutes to read a page of her book.

minutes	3	■	■	■	■	■
pages	1	2	3	4	5	6

1. How many pages can she read in 15 minutes?

2. At this rate, how many pages will she read in $\frac{1}{2}$ hour?

The cars on the freeway are traveling 55 miles per hour.

miles	55	■	■	■	■	■
hours	1	2	3	4	5	6

3. How far will they go in 5 hours?

4. About how long will it take to travel 300 miles?

Howard delivers 14 newspapers in 10 minutes.

papers	14	■	■	■	■	■
minutes	10	20	30	40	50	60

5. How many papers can Howard deliver in 1 hour?

6. How long will it take Howard to deliver the 70 papers on his route?

Tomatoes cost 65¢ a pound.

price	\$1.30	■	■	■	■	■
pounds	2	4	6	8	10	12

7. How much do 10 pounds of tomatoes cost?

8. About how many pounds of tomatoes can you buy with \$5.00?

Practice Set 78 (cont.)

Write all of your answers on a separate sheet of paper.

Rename the following fractions as decimals.

9. $\frac{8}{100}$ **10.** $\frac{6}{10}$ **11.** $\frac{3}{10}$ **12.** $\frac{1}{3}$

13. $\frac{250}{1,000}$ **14.** $\frac{47}{100}$ **15.** $\frac{4}{8}$ **16.** $\frac{1}{4}$

17. $\frac{6}{16}$ **18.** $\frac{42}{100}$ **19.** $\frac{182}{1,000}$ **20.** $\frac{14}{18}$

Write the next three numbers in the pattern.

21. 23,610; 23,615; 23,620

22. 39.55, 39.50, 39.45

23. 151, 148, 145

24. 1,455, 1,130, 805

Complete.

25. $10^2 =$ ■

26. $10^{■} = 1,000$

27. 10 * 10 * 10 * 10 * 10 * 10 = $10^{■}$

Volume = length × width × height
= area of base × height

Find the volume of each rectangular prism.

28.

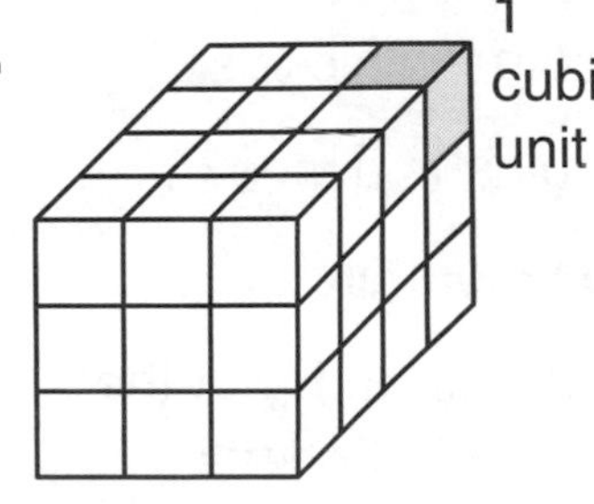

29.

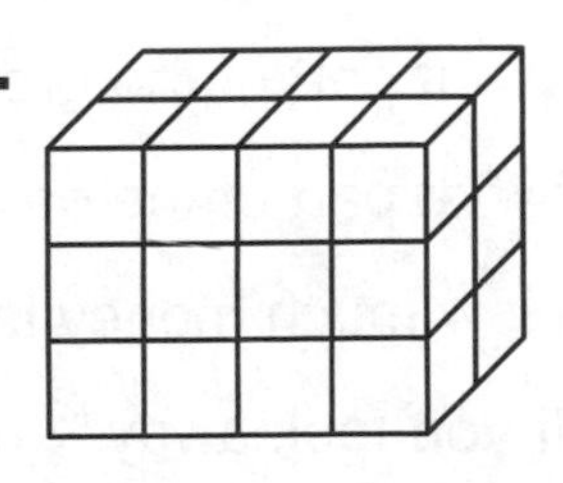

Practice Set 79

Write all of your answers on a separate sheet of paper.

Solve.

1. Mr. Turner drives 260 miles and uses 10 gallons of gas. How far could he drive on 1 gallon of gas?
2. Ms. Smith buys 3 pounds of apples for $2.04. What is the price for 1 pound of apples?
3. Allison sells 6 cards for $9.00. What is the price per card?
4. Tonya jogs 8 miles in 1 hour 12 minutes. What is her rate per mile?

Answer the following questions. If the answer is a fraction, write it in simplest terms.

5. What part of the group of coins is pennies?
6. What part of the group of coins is nickels?
7. What part of the group of coins is dimes?
8. How much money is in the whole group?
9. If you took away $\frac{2}{3}$ of the nickels and $\frac{5}{6}$ of the pennies, how much money would be left?

Practice Set 80

Write all of your answers on a separate sheet of paper.

Find the unit price for each. Tell which is the better buy.

1. a. 4 cans of peaches for $1.00
b. 2 cans for peaches for $0.60

2. a. 6 ounces of raisins for $1.68
b. 1 pound of raisins for $3.52

3. a. 8 eggs for $2.00
b. 1 dozen eggs for $2.40

4. a. 3 juice boxes for $1.05
b. 10 juice boxes for $3.50

Write <, > or = to make each number sentence true.

Reminder:	
>	greater than
<	less than
=	equal to

5. 47 + 63 ■ 22 + 74

6. 8 + 43 ■ 7 + 13 + 31

7. 85 + 23 ■ 81 + 35

8. 9 * 12 ■ 404 / 4

9. 169 − 40 ■ 95 + 26

10. Order these numbers from smallest to largest.

$\frac{1}{4}$ $\frac{3}{6}$ $\frac{1}{10}$ $\frac{9}{12}$ $\frac{16}{16}$

Write all of your answers on a separate sheet of paper.

Use the spinner for items 11–14. Suppose you spin a paper clip on the base of the spinner. Write *true* or *false* for each statement.

11. The paper clip is most likely to land on green.

12. The paper clip has an equal chance of landing on red or blue.

13. The paper clip is 2 times as likely to land on blue as on green.

14. The paper clip is more likely to land on blue than on yellow.

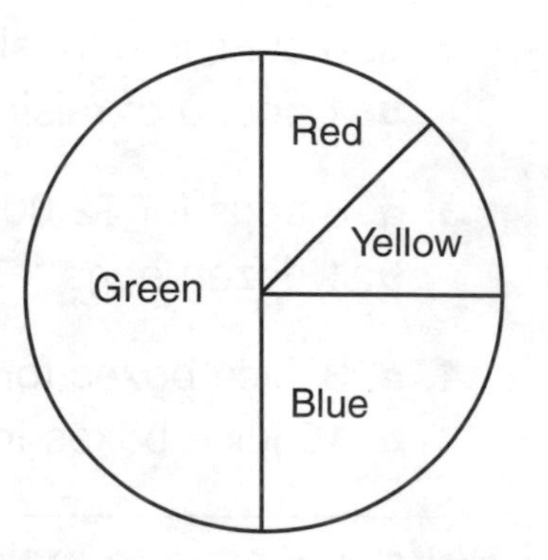

Complete.

15. If 10 counters are $\frac{1}{2}$, then ___ counters are the ONE.

16. If 3 counters are $\frac{1}{4}$, then ___ counters are the ONE.

17. If 8 counters are $\frac{2}{3}$, then ___ counters are the ONE.

18. If 70 counters are $\frac{7}{9}$, then ___ counters are the ONE.

Write *true* or *false* for each number sentence.

19. $14 + 13 = 27$

20. $6 * 9 = 48$

21. $4 * 7 < 30$

22. $41 - 25 = 18$

23. $3 * 3 = 54 \div 6$

24. $4 * (6 + 2) = 24$